MONOGRAPHS ON THE PHYSICS AND CHEMISTRY OF MATERIALS

MONOGRAPHS ON THE PHYSICS AND CHEMISTRY OF MATERIALS

Strong solids (Third edition) A. Kelly and N. H. Macmillan
Optical spectroscopy of inorganic solids B. Henderson and G. F. Imbusch
Quantum theory of collective phenomena G. L. Sewell
Experimental high-resolution electron microscopy (Second edition) J. C. H. Spence
Experimental techniques in low-temperature physics Guy K. White
Principles of dielectrics B. K. P. Scaife
Surface analytical techniques J. C. Rivière
Basic theory of surface states Sydney G. Davison and Maria Stęślicka
Acoustic microscopy Andrew Briggs
Dynamic light scattering W. Brown
Quasicrystals: a primer (Second edition) C. Janot
Interfaces in crystalline materials A. P. Sutton and R. W. Balluffi
Rare-earth iron permanent magnets J. M. D. Coey (ed.)

Basic Theory of Surface States

Sydney G. Davison

Departments of Applied Mathematics and Physics
University of Waterloo
Ontario, Canada

Maria Stęślicka

Institute of Experimental Physics
University of Wrocław
Wrocław, Poland

CLARENDON PRESS · OXFORD

This book has been printed digitally and produced in a standard specification in order to ensure its continuing availability

OXFORD
UNIVERSITY PRESS

Great Clarendon Street, Oxford OX2 6DP

Oxford University Press is a department of the University of Oxford.
It furthers the University's objective of excellence in research, scholarship,
and education by publishing worldwide in

Oxford New York

Auckland Cape Town Dar es Salaam Hong Kong Karachi
Kuala Lumpur Madrid Melbourne Mexico City Nairobi
New Delhi Shanghai Taipei Toronto
With offices in
Argentina Austria Brazil Chile Czech Republic France Greece
Guatemala Hungary Italy Japan South Korea Poland Portugal
Singapore Switzerland Thailand Turkey Ukraine Vietnam

Published in the United States
by Oxford University Press Inc., New York

Reprinted 2008

ISBN 978-0-19-851896-9

Printed and bound by CPI Antony Rowe, Eastbourne

To Prudence: a loving wife and a devoted friend.

Surfaces are not just here or there
For surfaces are everywhere
Splitting nature into phases
Solids, liquids and the gases.

S.G.D.

CONTENTS

In nature's infinite book of secrecy
A little can I read.
— William Shakespeare

PREFACE

Here the boundaries meet and all contradictions exist side by side.
— Fëdor Dostoevski

Surfaces are the boundaries between the different phases of matter and, as such, are the meeting places for a host of phenomena and processes at both the micro- and macro-scopic levels. From the study of this vast and complex world of surfaces has evolved the multi- and inter-disciplinary subject known as *Surface Science.* Of all the various phase boundaries, the gas-solid one is the most-widely studied, especially since the advent of ultra-high vacuum systems in the 1960's, which heralded in the golden age of surface science that we still enjoy today. These systems enable reproducible experiments to be performed, on well-characterized solid surfaces, under controlled, clean conditions.

In the voluminous literature, which has accumulated on the physics and chemistry of solid surfaces over the years, there are surprisingly few books. Moreover, virtually all of them stress the experimental point of view, though recently Zangwill (1988) adopted a more balanced approach by entwining theory with experiment in his *Physics at Surfaces.* On the theoretical side, we have the masterly monograph by Garcia-Moliner and Flores (1979) entitled *Introduction to the Theory of Solid Surfaces*, which is largely written in terms of the language of Green functions. By way of contrast, the present work follows a more pedestrian path, by postponing the treatment of Green-function methods until next to the last chapter. Thus, we have endeavoured to provide a 'primer' to the more advanced treatise and, in so doing, tried to make the theory of surface states more accessible to experimentalists and newcomers to the field.

Although the electronic theory of solids has traditionally been concerned with the motion of electrons in the bulk of the crystal, advances in this area have also fostered, albeit at a much slower rate, the development of the theory of surface states. The last two decades have, however, witnessed an ever-increasing interest in the behaviour of electrons at solid surfaces, which has stemmed from the technological importance of their role in solid-state devices and chemical reactions. Sometimes it is said that physicists look at a surface from inside the solid, while chemists view it from the outside. Here, the former standpoint is adopted, namely, the surface is regarded as the termination of the periodic lattice of an infinite crystal by a cleavage plane. In this way, the surface electronic properties may be discussed in terms of a perturbation of the crystal bulk properties. Thus, in order to study the electronic states associated with a surface, a knowledge of the bulk band structure is required. It is for this reason that Chapter 2, describing the necessary bulk properties, has been provided. Despite the leaning towards the physicist's point of view, it is hoped that the presentation of the subject matter will also appeal to physical chemists, and to the even broader audience of materials scientists and electronic engineers.

Because of the difficulties encountered in modelling the irregularities of real crystal surfaces, progress in surface-state theory has proceeded via the use of idealized models, though modern computers now make it possible to address complicated numerical surface problems. Nevertheless, simple models have provided a considerable amount of insight into the electronic structure of solid surfaces. Indeed, as P.W. Anderson (1978) pointed out in his Nobel Prize lecture, 'Very often a simplified model throws more light on the real workings of nature than any number of *ab initio* calculations of individual situations, which, even where correct, often contain so much detail as to conceal rather than reveal reality. It can be a disadvantage rather than an advantage to be able to compute or measure too accurately, since often what one measures or computes is irrelevant in terms of mechanism. After all, the perfect computation simply reproduces Nature, it does not explain her.' It is in the spirit of this philosophy that this book is written, since such a modellistic approach also provides the easiest and most pedagogical introduction to the theory of surface states. While an effort has been made to make the book as self-contained as possible, a knowledge of elementary quantum mechanics (Merzbacher 1970) and solid-state theory (Ashcroft and Mermin 1976) is assumed, so the material covered is intended for fourth-year undergraduate or first-year graduate students.

In searching for a means of selecting the topics to be included, the evolution of the theory of surface states suggested itself as the most logical theme, and also had the merit of marshalling the chapters into an appropriate order. Each chapter follows essentially the same 'recipe', starting with a general description of a particular method used to solve the Schrödinger equation, then applying it to a certain surface problem, which is defined in terms of a set of boundary conditions. By choosing a variety of problems, and analyzing them via different techniques, a fairly wide spectrum of surface-state studies is discussed, though no claim is made that the present treatment is at all comprehensive. On the contrary, we have only 'scratched the surface' of the subject. In this respect, the words that Francis Bacon wrote concerning his book on the *Advancement of Learning* seem fitting — 'We have contented ourselves to awake better spirits, like the bellringer, who is first up to call others to church.'

Let us now focus briefly on the chapters themselves. In electing to travel the historical route, it was natural to open the book with a tribute to Igor Tamm, the founder of the subject. We are therefore greatly indebted to D.A. Kirzhnits, from Tamm's own department, for providing the biographical sketch and photograph in Chapter 1. As was mentioned earlier, Chapter 2 contains the essential background material, on the electronic properties of solids, on which much of the subsequent chapters is based. Chapter 3 traces Tamm's pioneering steps that led him to the localization concept of surface states by a simple wave-function matching procedure. In the next chapter, the electron-orbital picture is used to introduce the classifications of Tamm and Shockley states according to the type of surface termination of the crystal. There then follows a discussion of the nearly-free electron approach in Chapter 5, where the idea of a complex wave number gives rise to the analytic continuation of both the energy and wave function across a forbidden energy gap. The Mathieu potential model of Chapter 6 provides a bridge between the tight-binding and nearly-free electron treatments of the previous chapters and, in this way, helps to unify the orbital and potential perspectives of surface states. Field effects are described in Chapter 7, where they manifest themselves in the forms of virtual and image states, which are very important in understanding field-emission and inverse-photoemission experiments, respectively. The most powerful and versatile techniques are the Green-function ones, encountered in Chapter 8, which enable the density of states (an essential quantity in interpreting spectroscopic data) to be calculated in a straightforward manner. Although these techniques are of a rather complicated mathematical nature, their presentation is made as simple as possible by providing the step-by-step detail used in deriving all the equations — a policy followed throughout the course of the book. The final chapter is devoted to an extensive description of the historical development of surface states, from the early work of the pioneers, to that of more recent times. A series of chronologically overlapping theoretical and experimental themes wind their way through the chapter, each one tracing the evolutionary thread of some particular topic, thus weaving a 'tapestry' depicting the story of surface states.

Quotations are found as sub-titles to the chapter headings, and are sprinkled elsewhere in the text. For, as Emerson observed, 'By necessity, by proclivity, and by delight, we all quote.' Indeed, quotations bring a certain ambience of charm to any work they grace, and a sense of literary relief in heavy tomes of scientific thought. Yet, much more than this, they express the very essence of the human experience, as distilled by the columns of time. In being literary windows on the scientific scene, their inclusion here reflects C.P. Snow's philosophy of the *Two Cultures — the Sciences and the Humanities*, since, as science lives in our minds, so poetry should dwell in our hearts.

It now remains for us to remember, with both pleasure and gratitude, all those who have helped us in writing this book. We are fortunate to have enjoyed the support of our friends and colleagues in Waterloo and Wrocław for many years. In particular, we are most indebted to R. Kucharczyk and K.W. Sulston for their critical reading of the manuscript, and to F.O. Goodman and W. Kisiel for their contagious enthusiasm and constant encouragement. We have also benefited greatly from our correspondence and visits with A.T. Amos, L. Dobrzynski, M.L. Glasser, W. Steinmann and H. Ueba. For the beautifully prepared computer-generated camera-ready manuscript, our thanks and admiration go to Lynda Clarke. We also thank Alan Hildebrand for his skillful rendition of our figures. Those reproduced from other papers are identified in their captions, and we are grateful to the authors and publishers for their permission to use them. The task of 'quality control' fell to the team of Ken Sulston and Pru Davison, whose patience and sharp eyes reduced the mistakes to a minimum, through seemingly endless proof-reading sessions. Lastly, we wish to express our appreciation to the staff of Oxford University Press for their guidance and assistance in preparing this work for publication.

Waterloo S.G.D.
Wrocław M.S.
March 1991

ABBREVIATIONS

It is when I struggle to be brief
that I become obscure.
— Horace

AO	Atomic orbital
DODS	Different orbitals for different spins
DOS	Density of states
FEG	Forbidden energy gap
GF	Green function
H	Hybridization
KP	Kronig-Penney
LCAO	Linear combination of atomic orbitals
LDOS	Local density of states
LEED	Low-energy electron diffraction
MO	Molecular orbital
NFE	Nearly-free electron
NH	Non-hybridization
SDOS	Surface density of states
SGF	Surface Green function
TBA	Tight-binding approximation

1

FOUNDING FATHER

Discovery consists of seeing what everybody has seen and thinking what nobody has thought.
— Albert Szent-Gyorgyi

1.1 Igor E. Tamm (1895-1971)

At the beginning of 1932, the journal *Physik Zeitschrift der Sowjetunion* published the paper 'On the possible bound states of electrons on a crystal surface' which indicated, for the first time, the existence of specific electron states (later to be called 'Tamm states') localized near the crystal boundary. Subsequently, this article played an important role in the development of the theory of surface and contact properties of solids and their numerous applications.

The author of the paper, Professor Igor Evgenievich Tamm, was one of the outstanding physicists of the twentieth century. He was not a narrow expert in solid-state physics but, on the contrary, always tried to participate in the elaboration of 'hot' physical problems of the time. The breadth of Tamm's interests, and the weight of his contribution to physics, may be easily seen from the following brief description of only some of his many projects, and also from the recently published selection of his papers (Bolotovskii and Frenkel 1990).

In 1930, Tamm formulated a quantum theory of light scattering in crystals on the basis of the concepts of elastic wave quanta (phonons). In particular, he described the so-called Raman scattering phenomenon, which had just been discovered simultaneously by his teacher Leonid Mandelstam and Grigory Landsberg and Chandrasekhar Raman in India. That same year, he made essentially the first consistent calculation of the quantum-electrodynamical effect involved in the scattering of a photon by an electron; the corresponding equation for the scattering cross-section is often termed the Klein-Nishina-Tamm formula. The following year, he put forward a quantum theory of the photo-electric effect in metals, which contained the basic idea for gaining insight into this phenomenon. In 1934, he proposed the exchange nature of nuclear forces, which led Hideki Yukawa to hypothesize the existence of the meson and its role as a carrier of nuclear interactions. During the period 1937-9, Tamm and Ilya Frank established the nature and developed the theory of Cerenkov radiation for which they were awarded the 1958 Nobel Prize. The Tamm-Dankov method for describing strong interactions, which became widely used in the theory of elementary particles, was formulated in 1945. In subsequent years, he became involved in the problem of magnetic confinement of hot plasmas for thermonuclear fusion.

Igor E. Tamm was born on June 8, 1895 in Vladivostok. His father was an engineer, and he spent his childhood and youth in the Ukraine. He received his education in physical science in Edinburgh and Moscow. On graduating from Moscow University in 1918, Tamm taught physics in southern Russia, before returning to Moscow in 1922. For many years he was Head of the Faculty of Theoretical Physics in Moscow University. He organized the Department of Theoretical Physics in the P.N. Lebedev Physical Institute in 1934, and was its head until the end of his life.

Tamm created a large school of theoretical physicists, many of whom became famous and created their own scientific schools. As a scientific leader he was neither stern nor dictatorial. His students were given a free hand, but their results were subjected to exacting and strict criticism. At the same time, if they did something really worthwhile, each of them would be aware of Tamm's close support.

The spectrum of Tamm's interests was very wide and went far beyond the bounds of physics. He exhibited a lively interest in biology, which he believed would one day take over physics' leading role in natural science. He also thought very seriously about the position of scientists in the stormy post-war world. He was an active participant in the Pugwash movement for peace and disarmament. He was an enthusiastic sportsman, being an alpinist until the last years of his active life.

As a person, Igor Tamm was a warm, friendly and wise man, who was always ready to help in any necessary and just cause. Although his life was not an easy one, he was always cheerful. Consequently, it is not surprising that those who knew him held him in great respect and loved him deeply.

The scientific achievements of Igor Tamm were recognized not only in the U.S.S.R., where he was elected an Academician and awarded the title of a Hero of Socialist Labour, but also throughout the world. Besides receiving the Nobel Prize, he was elected a member of many foreign academies and scientific societies. His friends in other countries included such world-renowned scientists as Niels Bohr, Paul Dirac and Paul Ehrenfest.

Despite becoming seriously ill and bedridden in 1967, Tamm continued to work, meet many people and maintain a lively interest in science until his death on April 12, 1971 at the age of 76.

D.A. Kirzhnits
P.N. Lebedev Institute of Physics
Moscow, U.S.S.R.

2

ELECTRONS IN INFINITE CRYSTALS

All finite things reveal infinitude.
— Theodore Roethke

2.1 Preliminaries

The behaviour of electrons in a crystal can be described, in principle, by solving the Schrödinger equation for all the electrons in the solid. We say 'in principle', because it is well-known that any attempt to calculate the intricate electronic properties of crystals requires the application of rather formidable theoretical methods. The mathematical difficulties arise from two sources: the complex geometry of real crystals and the complicated potential fields of real atoms. Thus, to make the problem more tractable, it is necessary to introduce some simplifying approximations.

The aim of this chapter is to present the basic methods for calculating the energy band structure of an electron moving in a crystal. Band theory is based on two fundamental approximations. The first assumes that the electronic and nuclear motions can be separated by invoking the *adiabatic approximation* of Born and Oppenheimer (1927). A second simplification is made by using the *one-electron approximation*: each of the electrons is assumed to move under the influence of a periodic potential due to the fixed positive ions, and the average constant field of all the other electrons. Thus, the discussion of the methods for calculating the energy levels in crystals will be based on the one-electron Schrödinger equation with a periodic potential. The fundamental theorem concerning the motion of electrons in a periodic potential is that due to Bloch (1928) which, in its one-dimensional form, is usually called the *Bloch-Floquet theorem.*

2.1.1 Bloch-Floquet theorem

Let the potential energy $V(x)$ be a periodic function of x of period a, so that

$$V(x) = V(x + na) \quad , \quad n = 0, \pm 1, \pm 2, ... \tag{2.1}$$

The one-dimensional motion of electrons of energy E in such a potential field is described by the Schrödinger equation

$$\psi''(x) + f(x)\psi(x) = 0 \tag{2.2}$$

where $f(x) = E - V(x)$, is also periodic in a, since E is a constant. Note, here and henceforth, *Rydberg's atomic units* are used, so that $\hbar = 2m = 1$. Solutions of (2.2) can be found with the help of a theorem which, in pure mathematics, is known as the Floquet (1883) theorem and, when applied to wave mechanics, is called *Bloch's theorem*. According to the Bloch-Floquet theorem, solutions of (2.2) may be written in the form

$$\psi(x) = \psi_k(x) = u_k(x)e^{ikx} \tag{2.3}$$

where $u_k(x)$ is a periodic function of period a and k is a constant to be determined by the boundary conditions. The solution (2.3) can also be expressed in another, frequently useful form. Suppose (2.2) is solved for the region $0 \leq x \leq a$, and the solution is given by

$$\psi(x) = \psi_0(x) \quad , \quad 0 \leq x \leq a. \tag{2.4}$$

Comparing (2.3) and (2.4) leads to

$$u_k(x) = \psi_0(x)e^{-ikx}. \tag{2.5}$$

Introducing into (2.3) the new coordinate

$$x = x' + na \quad ; \quad n = 0, \pm 1, \pm 2, ... \tag{2.6}$$

results in

$$\psi(x' + na) = u_k(x')e^{ikx'} \cdot e^{ikna} \tag{2.7}$$

because $u_k(x' + na) = u_k(x')$. Now, if $na \leq x \leq (n + 1)a$, then from (2.6) $0 \leq x' \leq a$ and, according to (2.5),

$$u_k(x') = \psi_0(x')e^{-ikx'}. \tag{2.8}$$

Inserting (2.8) and (2.6) into (2.7) gives

$$\psi(x) = \psi_0(x')e^{ikna} \quad , \quad na \leq x \leq (n+1)a. \tag{2.9}$$

The general solution of the wave equation (2.2) has the form

$$\psi(x) = Au_k(x)e^{ikx} + Bu_{-k}(x)e^{-ikx} \tag{2.10}$$

A and B being arbitrary constants. The constant k, called the *wave number*, is determined by the boundary conditions. If the periodic field extends over the entire range of x between $+\infty$ and $-\infty$, then, as is easily seen from (2.10), the wave function remains finite, if and only if, the wave number k is *real.*

2.1.2 Bloch theorem

The generalization of the Bloch-Floquet theorem to three dimensions is fairly obvious, and leads to the following form of the wave function

$$\psi_{\mathbf{k}}(\mathbf{r}) = u_{\mathbf{k}}(\mathbf{r})e^{i\mathbf{k}\cdot\mathbf{r}} \tag{2.11}$$

where $\mathbf{k}$ is the *wave vector* and the function $u_{\mathbf{k}}(\mathbf{r})$ has the three-dimensional periodicity of the crystal lattice, viz.,

$$u_{\mathbf{k}}(\mathbf{r}+\mathbf{a}_n) = u_{\mathbf{k}}(\mathbf{r}). \tag{2.12}$$

The vectors $\mathbf{a}_n$ may be expressed in terms of three primitive translation vectors $\mathbf{a}_1,\ \mathbf{a}_2,\ \mathbf{a}_3$ in the following way

$$\mathbf{a}_n = n_1\mathbf{a}_1 + n_2\mathbf{a}_2 + n_3\mathbf{a}_3 \tag{2.13}$$

where $n_1,\ n_2,\ n_3$ are integers.

Equation (2.11) was first derived by Bloch (1928), and is known as *Bloch's theorem.* Unfortunately, unlike the one-dimensional case, there is no simple analytical proof of this theorem. However, it is possible to demonstrate, from the periodicity of the electron potential energy, namely,

$$V(\mathbf{r}) = V(\mathbf{r}+\mathbf{a}_n) \tag{2.14}$$

that the electronic wave function has the form required by Bloch's theorem.

The Hamiltonian describing the motion of an electron in a solid can be written as

$$H = -\nabla^2 + V(\mathbf{r}) \tag{2.15}$$

where $V(\mathbf{r})$ satisfies (2.14). It is convenient to introduce the operator T_n of the translation by a lattice vector via the equation

$$T_n\psi(\mathbf{r}) = \psi(\mathbf{r} + \mathbf{a}_n). \tag{2.16}$$

These operators commute with each other, since

$$T_n T_{n'}\psi(\mathbf{r}) = \psi(\mathbf{r} + \mathbf{a}_{n'} + \mathbf{a}_n) = \psi(\mathbf{r} + \mathbf{a}_n + \mathbf{a}_{n'}) = T_{n'}T_n\psi(\mathbf{r}).$$

Clearly, when $V(\mathbf{r})$ is a function with the periodicity of the lattice, the translation operator commutes with the Hamiltonian, since it leaves H unchanged. Consequently, the translation operator T_n and the Hamiltonian H can have common eigenfunctions, and

$$T_n\psi(\mathbf{r}) = c_n\psi(\mathbf{r}). \tag{2.17}$$

The wave function $\psi(\mathbf{r})$ must also simultaneously satisfy the Schrödinger equation $H\psi(\mathbf{r}) = E\psi(\mathbf{r})$. Comparing (2.16) and (2.17) leads to

$$\psi(\mathbf{r} + \mathbf{a}_n) = c_n\psi(\mathbf{r}) \tag{2.18}$$

Now, if $\psi(\mathbf{r})$ is normalised for an infinite crystal, then

$$\int |\psi(\mathbf{r} + \mathbf{a}_n)|^2 d\mathbf{r} = \int |\psi(\mathbf{r})|^2 d\mathbf{r} \tag{2.19}$$

and, because of (2.18),

$$|c_n|^2 = 1 \tag{2.20}$$

so

$$c_n = e^{i\mathbf{k}\cdot\mathbf{a}_n}. \tag{2.21}$$

Equations (2.18) and (2.21) give

$$\psi(\mathbf{r} + \mathbf{a}_n) = e^{i\mathbf{k}\cdot\mathbf{a}_n}\psi(\mathbf{r}) \tag{2.22}$$

or

$$\psi(\mathbf{r}) = \psi(\mathbf{r} + \mathbf{a}_n)e^{-i\mathbf{k}\cdot\mathbf{a}_n} \tag{2.23}$$

so defining

$$u_{\mathbf{k}}(\mathbf{r}) = e^{-i\mathbf{k}\cdot(\mathbf{r}+\mathbf{a}_n)}\psi(\mathbf{r}+\mathbf{a}_n) \tag{2.24}$$

leads to

$$\psi(\mathbf{r}) = u_{\mathbf{k}}(\mathbf{r})e^{i\mathbf{k}\cdot\mathbf{r}}$$

which is the form required by Bloch's theorem. It only remains to show that the function $u_k(\mathbf{r})$ has the periodicity of the crystal lattice. From (2.24) comes

$$u_{\mathbf{k}}(\mathbf{r}+\mathbf{a}_m) = e^{-i\mathbf{k}\cdot(\mathbf{r}+\mathbf{a}_n+\mathbf{a}_m)}\psi(\mathbf{r}+\mathbf{a}_n+\mathbf{a}_m) \tag{2.25}$$

while (2.22) gives

$$\psi(\mathbf{r}+\mathbf{a}_n+\mathbf{a}_m) = e^{i\mathbf{k}\cdot\mathbf{a}_m}\psi(\mathbf{r}+\mathbf{a}_n). \tag{2.26}$$

Thus (2.26) in (2.25) yields

$$u_{\mathbf{k}}(\mathbf{r}+\mathbf{a}_m) = e^{-i\mathbf{k}\cdot(\mathbf{r}+\mathbf{a}_n)}\psi(\mathbf{r}+\mathbf{a}_n) = u_k(\mathbf{r}) \tag{2.27}$$

by (2.24), so the wave function of an electron in the periodic field of a crystal has the form of Bloch's function (2.11).

2.1.3 Reciprocal lattice and Brillouin zones

The reciprocal lattice is a geometrical construction, which enables the crystal geometry to be directly related to **k**-space, and is defined by means of the three basic vectors $\mathbf{b}_1$, $\mathbf{b}_2$, $\mathbf{b}_3$, which satisfy the equations

$$\mathbf{b}_i \cdot \mathbf{a}_j = \delta_{ij} \quad ; \quad i,j = 1,2,3 \tag{2.28}$$

where $\mathbf{a}_1$, $\mathbf{a}_2$, $\mathbf{a}_3$ are the three basic vectors of the direct lattice, and the reciprocal lattice vector

$$\mathbf{b}_{\mathbf{g}} = g_1\mathbf{b}_1 + g_2\mathbf{b}_2 + g_3\mathbf{b}_3 \tag{2.29}$$

g_1, g_2, g_3 being integers.

As can be seen from (2.21), the wave vector **k** is not uniquely determined in Bloch's function (2.11), because the replacement of **k** by $(\mathbf{k} + 2\pi\mathbf{b}_{\mathbf{g}})$ leaves the function unaffected, i.e.,

$$e^{i(\mathbf{k}+2\pi\mathbf{b}_{\mathbf{g}})\cdot\mathbf{a}_n} = e^{i\mathbf{k}\cdot\mathbf{a}_n}$$

since, from (2.13), (2.28) and (2.29),

$$e^{i2\pi \mathbf{b_g}\cdot\mathbf{a}_n} = e^{i2\pi\times integer} = 1.$$

With vectors **k** and $(\mathbf{k}+2\pi\mathbf{b_g})$ being equivalent, it is possible to restrict **k** to a single cell of the reciprocal lattice (with its linear dimensions multiplied by 2π). This cell is called the *first Brillouin zone* (Brillouin 1930, 1953; Jones 1962).

2.2 Kronig-Penney Model

The general features of the band structure of an electron propagating in a crystal can be obtained from very simple one-dimensional considerations, as was first demonstrated by Kronig and Penney (1931). In their model, a crystal potential field is represented by a linear array of rectangular barriers (Fig. 2.1) which, in the next simplifying step, are transformed into a chain of Dirac δ-functions (Fig. 2.2). The Kronig-Penney (KP) model is chosen for two reasons. First, for such models of a periodic field, the Schrödinger equation can be solved exactly. Secondly, the KP model underlies the classic work of Tamm, and many other papers on surface states.

To solve the Schrödinger equation for the KP potential, many different approaches have been adopted. Here, the two most frequently used in studying surface states are discussed, namely, the direct matching procedure and the Green-function technique.

2.2.1 Direct matching procedure

(i) *Classical Kronig-Penney model*

Kronig and Penney considered the periodic potential illustrated in Fig. 2.1, for which

$$\begin{aligned} &V = V_1 \quad \text{for} \quad -b < x \leq 0 \\ &V = 0 \quad \text{for} \quad 0 < x \leq a \\ &V[x+(a+b)] = V(x) \end{aligned} \tag{2.30}$$

The wave equation of the problem is given in (2.2), and the solutions must be of the form of Bloch functions (2.3). Substituting (2.3) into (2.2) imposes on u, in the range $-b < x \leq 0$, the condition

$$u'' + 2iku' - (k^2 + k_0^2)u = 0 \tag{2.31}$$

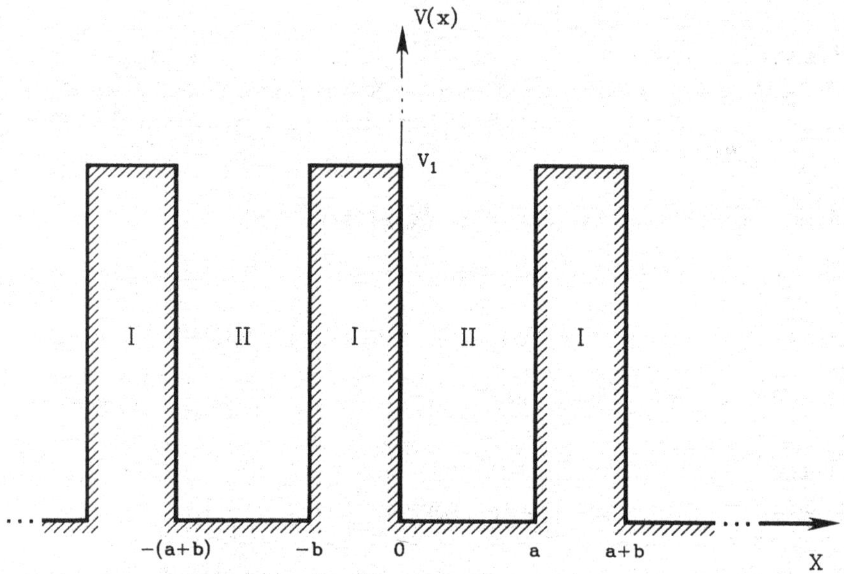

Fig. 2.1. Kronig-Penney model consisting of a periodic array of rectangular potential barriers of finite height (V_1) and width (b).

which has the solution

$$u = Ae^{(k_0 - ik)x} + Be^{-(k_0 + ik)x} \tag{2.32}$$

and in the region $0 < x \leq a$

$$u'' + 2iku' - (k^2 - k_1^2)u = 0 \tag{2.33}$$

whose solution is

$$u = Ce^{i(k_1 - k)x} + De^{-i(k_1 + k)x} \tag{2.34}$$

where

$$k_0 = (V_1 - E)^{1/2} \tag{2.35}$$

and

$$k_1 = E^{1/2} \tag{2.36}$$

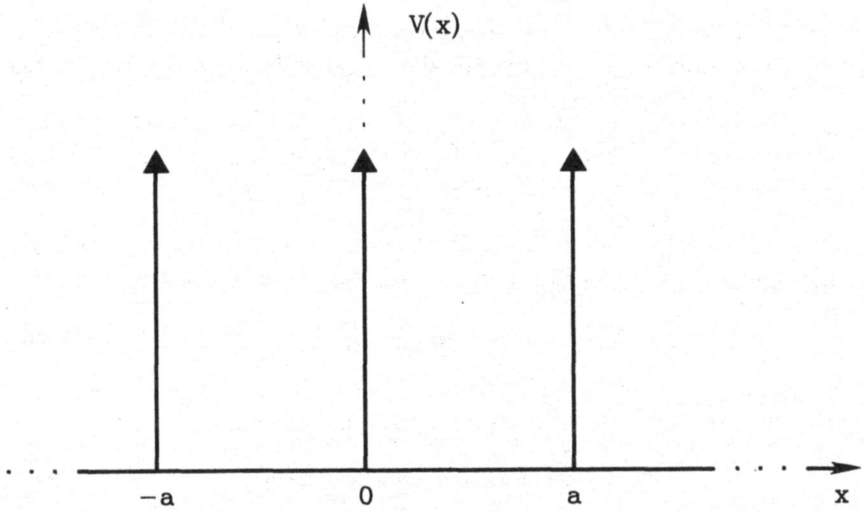

Fig. 2.2. *Dirac comb* δ-function representation of the Kronig-Penney potential.

may be assumed positive, without any loss of generality.

The continuity conditions across the potential interface between the regions I and II (Fig. 2.1) are

$$u_I(0) = u_{II}(0) \tag{2.37}$$

$$u_I'(0) = u_{II}'(0) \tag{2.38}$$

while the periodic nature of the function $u(x)$ means that

$$u_I(-b) = u_{II}(a) \tag{2.39}$$

$$u_I'(-b) = u_{II}'(a). \tag{2.40}$$

The four linear homogeneous equations resulting from these requirements are

$$A + B = C + D \tag{2.41}$$

$$(k_0 - ik)A - (k_0 + ik)B = i(k_1 - k)C - i(k_1 + k)D \tag{2.42}$$

$$Ae^{-(k_0-ik)b} + Be^{(k_0+ik)b} = Ce^{i(k_1-k)a} + De^{-i(k_1+k)a} \tag{2.43}$$

$$(k_0 - ik)Ae^{-(k_0-ik)b} - (k_0 + ik)Be^{(k_0+ik)b}$$
$$= i(k_1 - k)Ce^{i(k_1-k)a} - i(k_1 + k)De^{-i(k_1+k)a}. \tag{2.44}$$

For non-trivial solutions, the determinant of the coefficients of the constants must vanish. Expanding the determinant leads to the well-known KP relation for the energy, viz.,

$$\cos k(a + b) = \cosh k_0 b \cos k_1 a + [(k_0^2 - k_1^2)/2k_1 k_0] \sinh k_0 b \sin k_1 a. \tag{2.45}$$

(ii) *Kronig-Penney model with δ-barriers*

The KP relation (2.45) is rather complicated, so is more often used in its limiting form, i.e., after making the δ-potential transformation (Fig. 2.2)

$$\lim_{\substack{b \to 0 \\ V_1 \to \infty}} \tfrac{1}{2} k_0^2 ab = p \tag{2.46}$$

which converts (2.45) into the transcendental equation

$$\cos ka = \cos k_1 a + (p \sin k_1 a)/k_1 a. \tag{2.47}$$

A plot of the curve

$$\mathrm{g}(k_1 a) = \cos k_1 a + (p \sin k_1 a)/k_1 a \tag{2.48}$$

is given in Fig. 2.3. For each real value of k, $|\cos ka| \leq 1$. Thus, the solutions of (2.47) are given by the intersections of the straight lines in the range $-1 \leq \mathrm{g}(k_1 a) \leq 1$ and the curve (2.48). Projecting these intersections onto the $k_1 a$-axis shows that the energy values, which an electron moving through the lattice may have, form a spectrum consisting of solid lines (*allowed* energy bands) separated by finite intervals (*forbidden* energy gaps –FEG), where no solutions are found for real k-values.

The upper band edges are given by

$$k_1 a = (n + 1)\pi; \quad n = 0, 1, 2, \ldots \tag{2.49}$$

n being the band number, while the lower ones are obtained from (2.47), by setting $\cos ka = (-1)^n$, to give

$$k_1 a \tan(k_1 a/2) = p \ ; \quad n = 0, 2, 4 \ldots \tag{2.50}$$

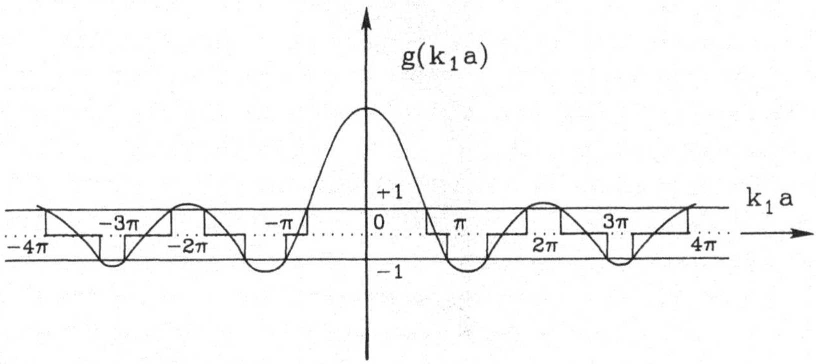

Fig. 2.3. Graphical solution of the Kronig-Penney dispersion relation (2.47).

$$k_1 a \cot(k_1 a/2) = -p \;\; ; \;\; n = 1, 3, 5, ... \tag{2.51}$$

It is of interest to study the influence of p on this spectrum. If $p = 0$, the curve of Fig. 2.3 becomes that of $\cos k_1 a$, the forbidden intervals on the $k_1 a$-axis disappear, and all the bands overlap to form a *continuum* of allowed levels from zero to infinity. This is the limiting case of *free electrons.* Increasing p, that is, increasing the binding energy of an electron to an individual atom, causes the appearance of the FEG and, thus, a decrease of the width of the allowed energy bands. When $p \to \infty$, (2.47) has solutions only if $\sin k_1 a = 0$. Therefore, the allowed bands reduce to the points $k_1 a = n\pi$ ($n = \pm 1, \; \pm 2, ...$), and the energy spectrum becomes *discrete*, i.e.,

$$E_n = (n\pi/a)^2 \;\; ; \quad n = 1, 2, 3, ... \tag{2.52}$$

which corresponds to an electron confined to a box of atomic dimensions with two impenetrable potential walls. Both cases are illustrated in Fig. 2.4, which shows the change in the energy spectrum during the transition from the free to the bound electron situation.

The explicit form of the wave function (2.3) for the KP potential can easily be found for the limiting case when $b \to 0$ (Fig. 2.1), since only the solutions in region II, namely (2.34), are now of interest. Thus,

$$\psi(x) = C(e^{ik_1 x} + \lambda e^{-ik_1 x}) \;\; , \;\; 0 < x \leq a \tag{2.53}$$

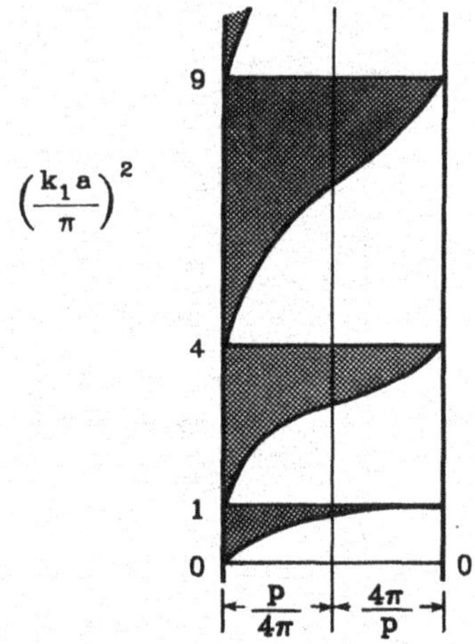

Fig. 2.4. Change in the energy spectrum during the transition from free to bound electrons in the δ-barrier Kronig-Penney model.

where $\lambda = D/C$ is obtained from (2.41) and (2.43), i.e.,

$$\lambda = [1 - e^{i(k_1-k)a}]/[e^{-i(k_1+k)a} - 1]. \tag{2.54}$$

The wave function for the cell extending from na to $(n+1)a$ can be found immediately from the Bloch-Floquet theorem (2.9).

It is worth noting that the classical KP model (Fig. 2.1), which has long remained largely a theoretical exercise, has been shown (Milanovic and Tjappin 1982, Cho and Prucnal 1988) to be a useful approximation for calculating the electronic energy level structure of semiconductor superlattices. A 'superlattice' is a *periodic structure* of thin layers of two materials (e.g., semiconductors) *in one-dimension* (cf. Esaki 1984 and references therein).

2.2.2 Green-function technique

(i) *Basic theory*

For a one-dimensional system, the Schrödinger equation can be written as

$$(\nabla^2 + k_1^2)\psi(x) = V(x)\psi(x) \tag{2.55}$$

where k_1 is given by (2.36). The solutions of (2.55) can be found by using the *Green function* of the operator of the left-hand side of (2.55), which is the operator for the motion of a free particle. The Green function of a free particle is defined as the function $G(x, y)$, which satisfies the equation for a point source, i.e.,

$$(\nabla^2 + k_1^2)G(x, y) = \delta(x - y). \tag{2.56}$$

Let $g(\ell, y)$ be the Fourier transform of $G(x, y)$, so that

$$G(x, y) = (2\pi)^{-1} \int_{-\infty}^{+\infty} g(\ell, y) e^{i\ell x} d\ell. \tag{2.57}$$

Inserting (2.57) in (2.56), and remembering that

$$\delta(x) = (2\pi)^{-1} \int_{-\infty}^{+\infty} e^{ikx} dk \tag{2.58}$$

yields

$$\int_{-\infty}^{+\infty} (k_1^2 - \ell^2) g(\ell, y) e^{i\ell x} d\ell = \int_{-\infty}^{+\infty} e^{i\ell(x-y)} d\ell. \tag{2.59}$$

Thus

$$g(\ell, y) = e^{-i\ell y}/(k_1^2 - \ell^2) \tag{2.60}$$

and (2.57) becomes

$$G(x, y) = (2\pi)^{-1} \int_{-\infty}^{+\infty} e^{i\ell(x-y)} (k_1^2 - \ell^2)^{-1} d\ell. \tag{2.61}$$

The integral in (2.61) can be evaluated by using the residue theorem (Arfken 1985, Priestley 1985). Its value, however, remains undefined until the form of the singularities at $\ell = \pm k_1$ is prescribed, subject to the boundary conditions being imposed upon $G(x, y)$, as $x \to \infty$. If $G(x, y)$ is taken to represent only *outgoing waves* at infinity, then we may choose the paths of integration shown in Fig. 2.5. The residue of the integrand at the pole, $\ell = \pm k_1$, which lies inside the contour of integration $\Gamma_\pm$, leads to the simple result

$$G(x, y) = -ie^{ik_1|x-y|}/2k_1. \tag{2.62}$$

The rule for avoiding the poles is found formally by replacing k_1 by $\pm(k_1 + i\epsilon)$ in the denominator of (2.61), where ϵ is a small quantity which, after the integral has been evaluated, is set equal to zero. For outgoing waves, ϵ is positive, so the poles of the integrand, $\ell = \pm(k_1 + i\epsilon)$, are shifted into the complex plane and again only one pole is inside each contour $\Gamma_{\pm}$.

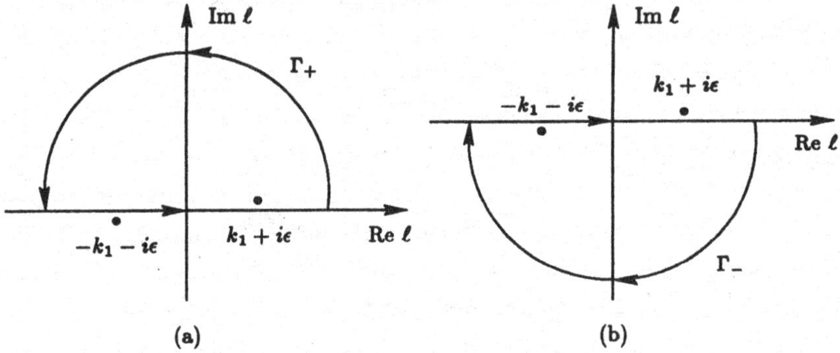

Fig. 2.5. Contours used in evaluating the Green function (2.61).

Having solved (2.56), its solution (2.62) is now used to assist in solving (2.55). Multiplying (2.55) by $G(x, y)$ and (2.56) by $\psi(y)$, integrating over all y and subtracting the resulting equations leads to

$$\int_{-\infty}^{+\infty} (\psi\nabla^2 G - G\nabla^2\psi)dy = \int_{-\infty}^{+\infty} \delta(x-y)\psi(y)dy$$

$$- \int_{-\infty}^{+\infty} V(y)\psi(y)G(x,y)dy. \tag{2.63}$$

Since

$$\int_{-\infty}^{+\infty} \delta(x-y)\psi(y)dy = \psi(x) \tag{2.64}$$

and the left-hand side of (2.63) is zero by Green's theorem (Morse and Feshbach 1953), it follows that the Schrödinger equation (2.55) can now be replaced by the integral equation

$$\psi(x) = \int_{-\infty}^{+\infty} V(y)\psi(y)G(x,y)dy. \tag{2.65}$$

(ii) *Periodic potential*

Let $V(y)$ in (2.65) be a periodic potential with period a. The integral in (2.65) can then be replaced by a sum of integrals over each unit cell, i.e.,

$$\psi(x) = \sum_{n=-\infty}^{+\infty} \int_{na}^{(n+1)a} V(y)\psi(y)G(x,y)dy. \tag{2.66}$$

Taking the crystal periodicity into account gives

$$V(y+na) = V(y)$$

while the Bloch-Floquet theorem (2.9) yields

$$\psi(y+na) = e^{ikna}\psi(y).$$

Thus, (2.66) can be written as

$$\psi(x) = \sum_{n=-\infty}^{\infty} e^{ikna} \int_0^a V(y)\psi(y)G(x,y+na)dy. \tag{2.67}$$

Denoting the integral part of x by $[x]$, and using (2.62) with $s = x - y$, enables the summation in (2.67) to be written as

$$\psi(x) = -\frac{i}{2k_1}\int_0^a \psi(y)V(y)\left[\sum_{n=-\infty}^{[s/a]} e^{ik_1 s}e^{i(k-k_1)na} + \sum_{n=[s/a]+1}^{\infty} e^{-ik_1 s}e^{i(k+k_1)na}\right] dy. \tag{2.68}$$

In summing the infinite geometrical series, it must be remembered that k_1 has a small positive imaginary part (cf. Fig. 2.5b), so that both series converge, and the result is

$$\psi(x) = -\frac{1}{2k_1}\int_0^a e^{ika[s/a]}\frac{e^{ika}\sin k_1 t - \sin k_1(t-a)}{\cos k_1 a - \cos ka}\psi(y)V(y)dy \tag{2.69}$$

where $t = s - a[s/a]$.

(iii) *Kronig-Penney potential*

The limiting case of the KP potential can be represented by a linear array of Dirac δ-functions (Fig. 2.2), i.e.,

$$V(y) = (2p/a) \sum_{n=-\infty}^{\infty} \delta(y - na) \tag{2.70}$$

where p is defined by (2.46) and is assumed to be positive. In the first cell of the crystal $(0 < y < a)$

$$V(y) = (2p/a)\delta(y - 0) \tag{2.71}$$

so (2.69) becomes

$$\psi(x) = -\frac{p}{k_1 a} \int_0^a e^{ika[s/a]}(\cos k_1 a - \cos ka)^{-1}$$

$$\times\{e^{ika} \sin k_1(x - a[x/a]) - \sin k_1(x - a[x/a] - a)\}\psi(y)\delta(y - 0)dy. \tag{2.72}$$

Putting $x' = x - a[x/a]$, and taking (2.64) into account, converts (2.72) into

$$\psi(x) = \frac{p}{k_1 a}\psi(0)e^{ik(x-x')}\frac{e^{ika} \sin k_1 x' - \sin k_1(x' - a)}{\cos ka - \cos k_1 a}. \tag{2.73}$$

The wave function $\psi(x)$ is now expressed in terms of its own value at the origin, located at one of the lattice sites. This leads at once to a relationship between the wave vectors k and k_1, which is proportional to the square root of the energy, since setting $x' = x = 0$ reduces (2.73) to the KP relation (2.47).

Note also that the wave function $\psi(x)$ is in the form required by Bloch's theorem, since it is periodic with the periodicity of the lattice, and represents a wave travelling to the right in the lattice. The replacement of k by $-k$ in (2.68) represents a wave travelling to the left. Thus, these two cases can be taken as the two independent solutions of the Schrödinger equation (Saxon and Hutner 1949).

Simple transformations show a complete equivalence between (2.73) and the wave function (2.53) obtained by Kronig and Penney. Indeed, introducing (2.54) into (2.53) gives

$$\psi(x) = C\left[e^{ik_1 x} + \frac{e^{-ik_1 x} \cdot e^{ika} - e^{-ik_1(x-a)}}{e^{-ik_1 a} - e^{ika}}\right] \tag{2.74}$$

which leads to

$$\psi(x) = 2iC\frac{e^{ika}\sin k_1x - \sin k_1(x-a)}{e^{ika} - e^{-ik_1a}} \tag{2.75}$$

so

$$\psi(0) = 2iC\frac{\sin k_1a}{e^{ika} - e^{-ik_1a}}$$

and (2.75) takes the form

$$\psi(x) = \frac{\psi(0)}{\sin k_1a}\left[e^{ika}\sin k_1x - \sin k_1(x-a)\right] \tag{2.76}$$

for the first crystal cell, i.e., for the range $0 < x < a$. However, for this range, (2.73) also takes the form of (2.76), since $x = x'$ and $\cos ka - \cos k_1a = p(\sin k_1a)/k_1a$ via the KP relation (2.47).

2.3 Crystal Orbital Method

The KP model is one of a very few examples of a periodic potential for which the Schrödinger equation can be solved exactly. In actual crystals, however, the potential function which is used must somehow be related to the actual potential experienced by an electron due to the ion cores and all the other electrons of the crystal. An exact solution of this problem, even in the one-electron approximation, is impossible; it is customary, therefore, to approach the problem from the viewpoint of either the *tight-binding approximation* (TBA) or the *nearly-free electron* (NFE) *approximation*, whichever seems the more appropriate to the particular situation.

The TBA, also known as the Hückel (1931) approximation in molecular orbital theory (Streitweiser 1966, Salem 1966), is valid whenever the motion of the valence electron is influenced to a much greater extent by the Coulomb potential of its parent ion than by all the other neighbouring ions. This is the case if the lattice ions are sufficiently far apart, or if the valence electrons are tightly bound to the parent ion.

In electron orbital theory (Coulson 1961), the crystal wave functions are based on the wave functions of the isolated atoms, called *atomic orbitals* (AO's). If the potential energy of an electron moving in the field of an isolated atom is $U(\mathbf{r})$, then the solutions of the Schrödinger equation

$$H_0\phi(\mathbf{r}) = E_0\phi(\mathbf{r}) \tag{2.77}$$

with

$$H_0 = -\nabla^2 + U(\mathbf{r}) \tag{2.78}$$

represent the AO's of the isolated atom, which correspond to the atomic energy levels E_0.

2.3.1 Simple cubic crystal

Consider a crystal of simple cubic structure, where the atoms are situated at points whose position vectors are

$$\mathbf{a_n} = n_1\mathbf{a}_1 + n_2\mathbf{a}_2 + n_3\mathbf{a}_3$$

$\mathbf{a}_1, \mathbf{a}_2, \mathbf{a}_3$ being three primitive translation vectors and n_1, n_2, n_3 integers. In the vicinity of the nth site, an electron in the isolated atom is described by the atomic orbital $\phi(\mathbf{r} - \mathbf{a_n})$. Since all sites are equivalent in an ideal crystal lattice, the electronic state with the energy E_0 is N-fold degenerate, N being the number of atoms in the crystal. As is known from perturbation theory, for degenerate levels (Davydov 1968), the zeroth-order wave function can be constructed as the *linear combination of atomic orbitals* (LCAO)

$$\psi(\mathbf{r}) = \sum_{\mathbf{n}} c_{\mathbf{n}}\phi(\mathbf{r} - \mathbf{a_n}). \tag{2.79}$$

The wave-function coefficients $c_{\mathbf{n}}$ can be found in a standard way (Davydov 1968), or determined from the requirement that the wave function (2.79) has the form of the Bloch functions (2.11). This condition is satisfied, if

$$c_{\mathbf{n}} = Ae^{i\mathbf{k}\cdot\mathbf{a_n}} \tag{2.80}$$

A being a constant. Indeed, inserting (2.80) into (2.79) gives

$$\psi(\mathbf{r}) = A\sum_{\mathbf{n}} e^{i\mathbf{k}\cdot\mathbf{a_n}}\phi(\mathbf{r} - \mathbf{a_n}) \tag{2.81}$$

which can be rewritten as

$$\psi(\mathbf{r}) = Ae^{i\mathbf{k}\cdot\mathbf{r}}\sum_{\mathbf{n}} e^{i\mathbf{k}\cdot(\mathbf{a_n}-\mathbf{r})}\phi(\mathbf{r} - \mathbf{a_n}) \tag{2.82}$$

i.e., the term

$$A\sum_{\mathbf{n}} e^{i\mathbf{k}\cdot(\mathbf{a_n}-\mathbf{r})}\phi(\mathbf{r} - \mathbf{a_n}) \tag{2.83}$$

plays the role of the function $u_{\mathbf{k}}(\mathbf{r})$ in (2.11). Consequently, (2.83) must also have the periodicity of the crystal lattice and, in fact, replacing $\mathbf{r}$ by $\mathbf{r}+\mathbf{a_m}$ in (2.83) leads to

$$A\sum_{\mathbf{n}} e^{i\mathbf{k}\cdot(\mathbf{a_n}-\mathbf{a_m}-\mathbf{r})}\phi(\mathbf{r}+\mathbf{a_m}-\mathbf{a_n}). \tag{2.84}$$

Setting $\mathbf{a_n}-\mathbf{a_m}=\mathbf{a_l}$ and changing the summation in (2.84) gives

$$A\sum_{\mathbf{l}} e^{i\mathbf{k}\cdot(\mathbf{a_l}-\mathbf{r})}\phi(\mathbf{r}-\mathbf{a_l}) \tag{2.85}$$

which obviously corresponds to the expression (2.83). Consequently, the wave function (2.83) satisfies the condition of translational symmetry, as required by Bloch's theorem, and thus, it can be used as a trial wave function for an estimate of the energy of an electron in a crystal. To be more specific, the TBA approach only allows the change in the energy of a valence electron to be found when the isolated atoms are brought together to form a crystal.

The wave function (2.81) must satisfy the Schrödinger equation for the entire crystal, i.e.,

$$H\psi(\mathbf{r})=E\psi(\mathbf{r}) \tag{2.86}$$

where

$$H=-\nabla^2+V(\mathbf{r}). \tag{2.87}$$

Obviously, the self-consistent potential

$$V(\mathbf{r})\neq\sum_{\mathbf{n}}U(\mathbf{r}-\mathbf{a_n}) \tag{2.88}$$

in (2.87).

Now, inserting (2.81) in (2.86), multiplying through on the left by $\phi^*(\mathbf{r})$ and integrating over all space, leads to

$$A\sum_{\mathbf{n}} e^{i\mathbf{k}\cdot\mathbf{a_n}}\left[\int\phi^*(\mathbf{r})H\phi(\mathbf{r}-\mathbf{a_n})d\tau-E\int\phi^*(\mathbf{r})\phi(\mathbf{r}-\mathbf{a_n})d\tau\right]=0. \tag{2.89}$$

However, from (2.87) and (2.78),

$$H=H_0+V(\mathbf{r})-U(\mathbf{r}). \tag{2.90}$$

Because of (2.90) and (2.77), the equation (2.89) takes the form

$$A\sum_{\mathbf{n}} e^{i\mathbf{k}\cdot\mathbf{a_n}}\left\{(E_0 - E)\int \phi^*(\mathbf{r})\phi(\mathbf{r}-\mathbf{a_n})d\tau + \int \phi^*(\mathbf{r})[V(\mathbf{r}) - U(\mathbf{r})]\phi(\mathbf{r}-\mathbf{a_n})d\tau\right\} = 0 \tag{2.91}$$

(i) *Atomic s-states*

If there is not too much overlapping of the AO's in the neighbouring cells, it is justified to assume that

$$\int \phi^*(\mathbf{r}-\mathbf{a_n})\phi(\mathbf{r}-\mathbf{a_m})d\tau = \delta_{\mathbf{nm}} \tag{2.92}$$

and to neglect all the terms beyond the *nearest-neighbour* contributions in (2.91). All the nearest-neighbour contributions will be the same, if the AO's $\phi(\mathbf{r})$ are spherically symmetric. This assumption therefore restricts our treatment to the case where the ground-state electronic configuration in the isolated atom is that of the s-state. However, there is no loss of generality in considering the s-states, since the extension to p-states is quite straightforward.

It is convenient to define the *Coulomb integral*

$$\alpha_s = \int \phi_s^*(\mathbf{r})[V(\mathbf{r}) - U(\mathbf{r})]\phi_s(\mathbf{r})d\tau \tag{2.93}$$

and the *resonance integral*

$$\beta_s = \int \phi_s^*(\mathbf{r})[V(\mathbf{r}) - U(\mathbf{r})]\phi_s(\mathbf{r}-\mathbf{a}_{\mathbf{n}_0})d\tau \tag{2.94}$$

where $\mathbf{a}_{\mathbf{n}_0}$ is a vector connecting the atom at the origin with the nearest neighbour atoms.

The Coulomb and resonance integrals are often defined in a slightly different way, i.e.,

$$\alpha = \int \phi^*(\mathbf{r})H\phi(\mathbf{r})d\tau \tag{2.93a}$$

$$\beta = \int \phi^*(\mathbf{r})H\phi(\mathbf{r}-\mathbf{a}_n)d\tau \tag{2.94a}$$

However, because of (2.90), (2.77) and (2.92),

$$\int \phi^* H\phi d\tau - \int \phi^*(V-U)\phi d\tau = E_0$$

where E_0 is the atomic energy level. Consequently, the difference between the definitions (2.93), (2.94) and (2.93a), (2.94a) is merely one of adjustment in the zero of energy.

Since $[V(\mathbf{r}) - U(\mathbf{r})] < 0$, to satisfy the minimum energy requirement in forming the crystal from the isolated atoms, the Coulomb integral (2.93) is always negative. The sign of the resonance integral (2.94) depends on the signs of the overlapping parts of the wave functions in (2.94). If these parts have the same sign, which is, for instance, the case for the s-states of the interior electrons, then $\beta_s < 0$ also.

In the case of a simple cubic crystal, each atom has six nearest neighbours. Inserting (2.92) to (2.94) in (2.91) leads to the following simple expression

$$E_0^s - E + \alpha_s + \beta_s \sum_{\mathbf{n}_0=1}^{6} e^{i\mathbf{k}\cdot\mathbf{a}\mathbf{n}_0} = 0. \tag{2.95}$$

Directing the axes x, y, z along the edges of a cube leads to

$$\sum_{\mathbf{n}_0=1}^{6} e^{i\mathbf{k}\cdot\mathbf{a}\mathbf{n}_0} = 2(\cos k_x a + \cos k_y a + \cos k_z a)$$

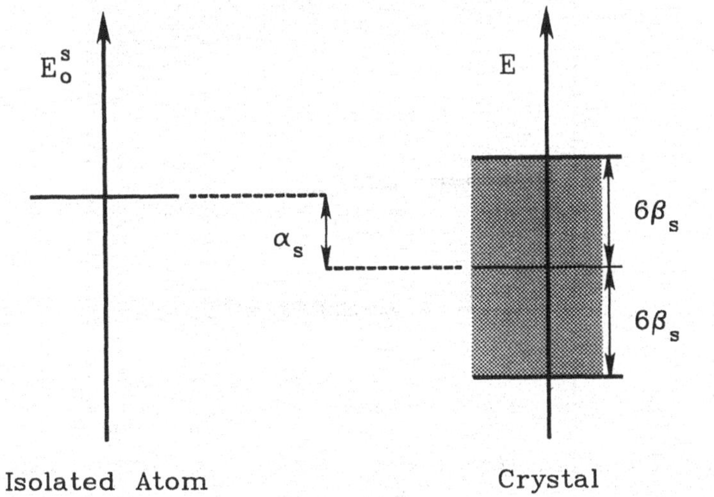

Fig. 2.6. Allowed energy s-band of a crystal formed by the interaction of coalescing isolated atomic states.

and, therefore, (2.95) gives

$$E = E_0^s + \alpha_s + 2\beta_s(\cos k_x a + \cos k_y a + \cos k_z a) \tag{2.96}$$

where a is the lattice constant.

Equation (2.96) gives the energy E of an electron in a crystal, obtained from the atomic energy level E_0^s. The minimum value of the energy is at the point $k_x = k_y = k_z = 0$ (assuming $\beta_s < 0$), i.e.,

$$E_{min} = E_0^s + \alpha_s + 6\beta_s. \tag{2.97}$$

The maximum values occur at $k_x = k_y = k_z = \pm\pi/a$, i.e.,

$$E_{max} = E_0^s + \alpha_s - 6\beta_s. \tag{2.98}$$

If $\beta_s > 0$, (2.97) corresponds to the maximum and (2.98) to the minimum of the energy E. The difference in energy between the maximum and minimum values is $12\beta_s$, which is the width of the *allowed energy band* corresponding to the s-state of the isolated atom (Fig. 2.6).

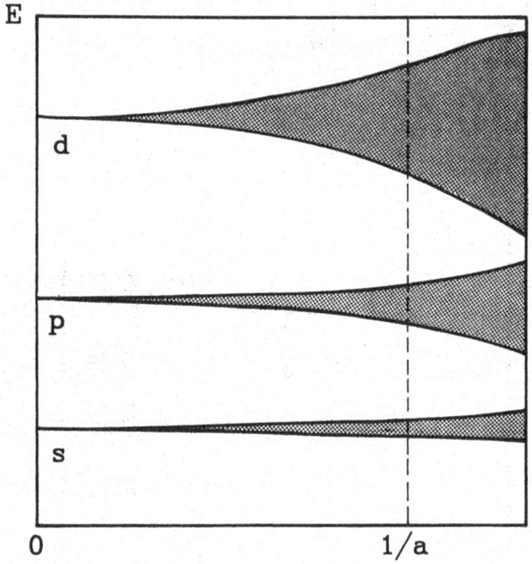

Fig. 2.7. Broadening of atomic s-, p- and d-levels into corresponding bands of finite width.

Just as the s-states of the individual atoms of a crystal become a band of electronic levels for the crystal as a whole, so do the p-states, d-states, etc., interact to give bands of electronic levels (Fig. 2.7). Consequently, the energy spectrum of an electron in a crystal is qualitatively similar to that obtained from the KP relation (2.47) (cf. also Figs. 2.3 and 2.4).

(ii) *Atomic p-states*

In the theory of surface states of semiconductors, atomic s- and p-states are of most interest, since they give rise to the valence and conduction bands in many semiconductors, including silicon and germanium. Thus, a brief discussion will be now given of the results of the interaction of the atomic p-states.

The wave functions of the stationary states of an electron in an arbitrary spherically symmetric field have the form (Davydov 1968)

$$\psi_{n\ell m}(r,\theta,\phi) = R_{n\ell}(r)Y_{\ell m}(\theta,\phi) \tag{2.99}$$

where $R_{n\ell}(r)$ is the radial wave function, depending on the principle quantum number n and the orbital quantum number $\ell \leq n-1$, and $Y_{\ell m}(\theta,\phi)$ are the spherical harmonics, depending on ℓ and the magnetic quantum number $m = 0,\ \pm 1,\ \pm 2,\ \ldots \pm \ell$. Now, the states with $\ell = 1$ are called p-states (s-states are those with zero orbital angular momentum, $\ell = 0$). Since the energy of an electron depends only on the quantum numbers n and ℓ, the p-state is characterized by three degenerate wave functions with $m = 0,\ +1,\ -1$, namely,

$$\psi_{n10} = R_{n1}(r)(3/4\pi)^{1/2}\cos\theta \tag{2.100a}$$

$$\psi_{n11} = R_{n1}(r)(3/8\pi)^{1/2}\sin\theta\ e^{i\phi} \tag{2.100b}$$

$$\psi_{n1,-1} = R_{n1}(r)(3/8\pi)^{1/2}\sin\theta\ e^{-i\phi}. \tag{2.100c}$$

Obviously, any linear combination of these functions will also give the wave function of a p-state. It is convenient to choose the following three linear combinations of (2.100)

$$\tfrac{1}{2}\left(\psi_{n11} + \psi_{n1,-1}\right) = r\sin\theta\cos\phi\cdot f(r) \equiv \psi_x \tag{2.101a}$$

$$\tfrac{1}{2i}\left(\psi_{n11} - \psi_{n1,-1}\right) = r\sin\theta\sin\phi\cdot f(r) \equiv \psi_y \tag{2.101b}$$

$$\psi_{n10} = r\cos\theta\cdot f(r) \equiv \psi_z \tag{2.101c}$$

where $f(r)$ is the radial wave function, which is determined by the form of the potential energy $U(\mathbf{r})$.

Because of (2.81) and (2.101a) to (2.101c), the wave function of an electron in the crystal can now be written as

$$\psi(\mathbf{r}) = \sum_{\mathbf{n}} e^{i\mathbf{k}\cdot\mathbf{a_n}} \left[d_1\psi_x(\mathbf{r}-\mathbf{a_n}) + d_2\psi_y(\mathbf{r}-\mathbf{a_n}) + d_3\psi_z(\mathbf{r}-\mathbf{a_n})\right] \quad (2.102)$$

where the coefficients d_1, d_2 and d_3 are determined from degenerate perturbation theory. However, for a simple cubic lattice, states of different types do not combine (Mott and Jones 1958), so, only one of the coefficients d_1, d_2 and d_3 can be different from zero. This implies that the relations corresponding to (2.91) separate into three groups involving the ψ_x, ψ_y or ψ_z, respectively, which accordingly may be treated independently. The treatment runs parallel to that for s-states and leads to the following expressions for the energy of an electron in the crystal: for a p_x-state (i.e., for $d_1 \neq 0$ and $d_2 = d_3 = 0$)

$$E_x = E_0^p + \alpha_p + 2\beta_p \cos k_x a + 2\gamma_p(\cos k_y a + \cos k_z a) \quad (2.103)$$

for a p_y-state (i.e., for $d_2 \neq 0$ and $d_1 = d_3 = 0$)

$$E_y = E_0^p + \alpha_p + 2\beta_p \cos k_y a + 2\gamma_p(\cos k_x a + \cos k_z a) \quad (2.104)$$

for a p_z-state (i.e., for $d_3 \neq 0$ and $d_1 = d_2 = 0$)

$$E_z = E_0^p + \alpha_p + 2\beta_p \cos k_z a + 2\gamma_p(\cos k_x a + \cos k_y a). \quad (2.105)$$

Here, E_0^p is the energy of a p-electron in an isolated atom and $\alpha_p, \beta_p, \gamma_p$ are the corresponding integrals for p-states, i.e.,

$$\alpha_p = \int \psi_x^*(\mathbf{r})[V(\mathbf{r}) - U(\mathbf{r})]\psi_x(\mathbf{r})d\tau \quad (2.106)$$

and

$$\int \psi_x^*(\mathbf{r})[V(\mathbf{r}) - U(\mathbf{r})]\psi_x(\mathbf{r}-\mathbf{a_{n_0}})d\tau$$

$$= \begin{cases} \beta_p & \text{for the nearest neighbours on the x-axis} \\ \gamma_p & \text{for the nearest neighbours on the y- or z-axis.} \end{cases} \quad (2.107)$$

Because of cubic symmetry, replacing ψ_x by ψ_y or ψ_z does not change the integrals (2.106) and (2.107).

(iii) *Hybridization of s- and p-orbitals*

If the s- and p-states lie close enough to one another, then their mutual interaction becomes important and must be taken into account. For simplicity, the treatment will be restricted to p_x-states only. Consider, therefore, a crystal lattice in which both an s-wave function (ϕ_s) and p_x-wave function (ϕ_p) are assigned to each lattice site; $\phi_s(\mathbf{r}-\mathbf{a_n})$ are symmetrical to the reflection across the lattice sites and $\phi_p(\mathbf{r}-\mathbf{a_n})$ antisymmetrical to such an operation. The general LCAO expansion is given by (2.79). Now, however, the p_x-wave functions must also be included, which leads to

$$\psi(\mathbf{r}) = \sum_{\mathbf{n}} [c_{\mathbf{n}}\phi_s(\mathbf{r}-\mathbf{a_n}) + d_{\mathbf{n}}\phi_p(\mathbf{r}-\mathbf{a_n})]. \tag{2.108}$$

The wave-function coefficients in (2.108) can be written as (cf. 2.80)

$$c_{\mathbf{n}} = Ae^{ia(n_1k_x+n_2k_y+n_3k_z)} \tag{2.109}$$

$$d_{\mathbf{n}} = Be^{ia(n_1k_x+n_2k_y+n_3k_z)} \tag{2.110}$$

where A and B are constants and n_1, n_2, n_3 are integers. Substituting (2.108) to (2.110) into the Schrödinger equation (2.86), and neglecting the interaction of all but the nearest neighbours, we obtain, in a manner completely analogous to the derivation for the s-states, the following homogeneous equations for the constants A and B

$$A[E - E_0^s - \alpha_s - 2\beta_s(\cos k_xa + \cos k_ya + \cos k_za)] - 2Bi\beta_{sp}\sin k_xa = 0$$

$$\begin{aligned} 2Ai\beta_{sp}\sin k_xa + B[E - E_0^p - \alpha_p - 2\beta_p\cos k_xa \\ -2\gamma_p(\cos k_ya + \cos k_za)] = 0. \end{aligned} \tag{2.111}$$

The new quantity β_{sp} is given by the formula

$$\begin{aligned} \int \phi_s^*(\mathbf{r}-\mathbf{a_{n'}})[V(\mathbf{r}) - U(\mathbf{r})]\phi_p(\mathbf{r}-\mathbf{a_n})d\tau \\ = \beta_{sp}(\delta_{\mathbf{n}_1',\mathbf{n}_1+1} - \delta_{\mathbf{n}_1',\mathbf{n}_1-1}). \end{aligned} \tag{2.112}$$

All other integrals are defined by (2.93) and (2.94) for s-states and by (2.106) and (2.107) for p-states. Equation (2.111) was obtained under the assumption that all the wave functions are orthonormal (cf. 2.92).

Eliminating A and B from (2.111) leads to the following expression for the energy of an electron in a crystal, when the s- and p-states are degenerate, namely,

$$E = \tfrac{1}{2}(K_s + K_p) + (\beta_s + \beta_p)\cos k_x a \pm \tfrac{1}{2}\left\{[K_s - K_p + 2(\beta_s - \beta_p)\cos k_x a]^2 + 16\beta_{sp}^2 \sin^2 k_x a\right\}^{1/2}. \quad (2.113)$$

The upper (lower) sign refers to the upper (lower) band. The quantities K_s and K_p in (2.113) are defined as

$$K_s = E_0^s + \alpha_s + 2\beta_s(\cos k_y a + \cos k_z a) \quad (2.114)$$

$$K_p = E_0^p + \alpha_p + 2\gamma_p(\cos k_y a + \cos k_z a). \quad (2.115)$$

The energy expression (2.113) will take a somewhat simpler form, if we introduce the simplifying assumptions originally suggested by Artmann (1952). Since β_s, β_p and β_{sp} are all of comparable magnitude (i.e., $\sim 1eV$), and the sign of β_p is opposite to that of β_s and β_{sp}, it is justified, and convenient, to assume that

$$-\beta_s = -\beta_{sp} = \beta_p = 1. \quad (2.116)$$

Obviously, the energy (2.113) will now be expressed in the units of β and can be treated as a dimensionless quantity. Because of (2.116), expression (2.113) reduces to

$$E = \tfrac{1}{2}(K_s + K_p) \pm 2\left[\left(\frac{K_s - K_p}{4}\right)^2 - \tfrac{1}{2}(K_s - K_p)\cos k_x a + 1\right]^{\frac{1}{2}} \quad (2.117)$$

which can be transformed into

$$E = \tfrac{1}{2}(K_s + K_p) \pm 2\left[\left(1 - he^{ik_x a}\right)\left(1 - he^{-ik_x a}\right)\right]^{\frac{1}{2}} \quad (2.118)$$

where

$$h = \tfrac{1}{4}(K_s - K_p) = \tfrac{1}{4}(E_0^s - E_0^p) + \tfrac{1}{4}(\alpha_s - \alpha_p) - \tfrac{1}{2}(1 + \gamma_p)(\cos k_y a + \cos k_z a). \quad (2.119)$$

For the values of $k_x a$ lying in the interval $(0, \pm\pi)$, the minimum of the upper band for fixed (k_y, k_z) is given by

$$E_{min} = \tfrac{1}{2}(K_s + K_p) + 2|1 - |h|| \quad (2.120)$$

while the maximum of the lower band

$$E_{max} = \tfrac{1}{2}(K_s + K_p) - 2|1 - |h||. \quad (2.121)$$

2.3.2 Cyclic crystal

An infinite crystal can be represented by a finite one by, either allowing the number of atoms N to go to infinity, or imposing periodic boundary conditions at each end. In one-dimension, another way is to construct a *cyclic* crystal by joining the ends to form a circle (Fig. 2.8). Here the N atoms are numbered from 0 to $(N-1)$. The crystal wave function for the kth state $\psi_k(x)$ can be expanded as a LCAO's ϕ_n to give (cf. 2.79)

$$\psi_k = \sum_{n=0}^{N-1} c_{nk}\phi_n. \tag{2.122}$$

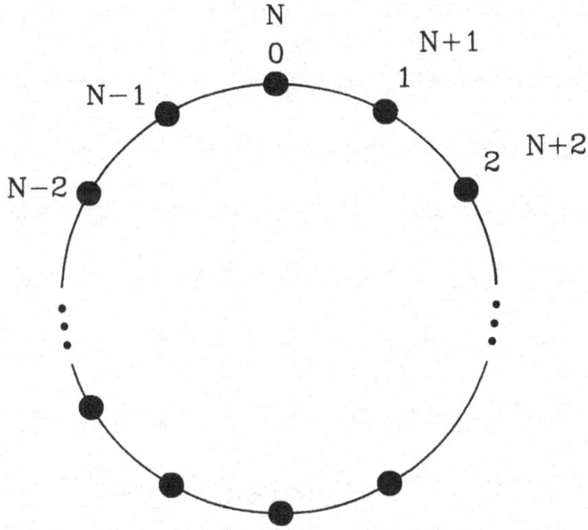

Fig. 2.8. Cyclic crystal of N atoms, for which $c_{0k} = c_{Nk}$.

In principle, the procedure for obtaining the energy expression for a cyclic crystal runs parallel to that for the simple cubic one. Therefore, inserting (2.122) into the Schrödinger equation

$$H\psi_k = E_k\psi_k \tag{2.123}$$

multiplying through on the left by ϕ_m^* and integrating over all space, leads to a system of homogeneous equations for c_{nk} (the so-called *secular equation*), namely,

$$\sum_n \left(\int \phi_m^* H \phi_n d\tau - E_k \int \phi_m^* \phi_n d\tau \right) c_{nk} = 0. \tag{2.124}$$

The solution of (2.124) is sought by assuming that the AO's form an orthonormal set (cf. 2.92) and neglecting all terms beyond the nearest-neighbour contributions (i.e., the TBA). Equation (2.124) now reduces to

$$(E_k - \alpha)c_{mk} = \beta(c_{m+1,k} + c_{m-1,k}) \tag{2.125}$$

where, according to (2.93a) and (2.94a),

$$\int \phi_m^* H \phi_n d\tau = \begin{cases} \alpha & \text{for } n = m \\ \beta & \text{for } n = m \pm 1. \end{cases} \tag{2.126}$$

Thus, the LCAO-TBA approach enables the Schrödinger equation to be cast in the form of a second-order difference equation with constant coefficients, whose solutions are found by setting (Hilderbrand 1968)

$$c_{mk} = x^m. \tag{2.127}$$

Substituting (2.127) in (2.125) leads to the quadratic equation

$$x^2 - 2X_k x + 1 = 0 \tag{2.128}$$

where

$$X_k = (E_k - \alpha)/2\beta \tag{2.129}$$

is the *dimensionless reduced energy.* Introducing the parameter θ_k, by choosing

$$X_k = \cos\theta_k \tag{2.130}$$

the positive root of (2.128) gives

$$x = e^{i\theta_k} \tag{2.131}$$

via Euler's formula. Hence, by (2.127), the solution of (2.125) is

$$c_{mk} = Ae^{im\theta_k}. \tag{2.132}$$

Since the wave function (2.122), with c_{mk} given by (2.132), must have the form required by the Bloch-Floquet theorem (2.9),

$$\theta_k \equiv ka. \tag{2.133}$$

i.e., θ_k plays the role of the wave number k.

The constant A in (2.132) can be found from the normalization condition

$$\int |\psi_k|^2 d\tau = 1. \tag{2.134}$$

Thus, with the aid of (2.132), substituting (2.122) into (2.134) yields

$$\sum_{n=0}^{N-1} A^2 = 1 \tag{2.135}$$

since $\int |\phi_m|^2 d\tau = 1$, so

$$A = N^{-\frac{1}{2}}. \tag{2.136}$$

Using (2.136) and (2.132), the cyclic-crystal wave function (2.122) becomes

$$\psi_k = N^{-\frac{1}{2}} \sum_{n=0}^{N-1} e^{in\theta_k} \phi_n. \tag{2.137}$$

It still remains to find the values of θ_k. These are obtained by inserting (2.132) in the Born-von Kármán (1912, 1913) *cyclic* boundary condition (Fig. 2.8) (Ziman 1965)

$$c_{0k} = c_{Nk} \tag{2.138}$$

which, on equating the real or imaginary parts, gives

$$\cos N\theta_k = 1 \quad \text{and} \quad \sin N\theta_k = 0 \tag{2.139}$$

i.e.,

$$\theta_k = 2\pi k/N \ ; \ \ k = 0, 1, \ldots \ N - 1. \tag{2.140}$$

For each value of θ_k, there is a corresponding energy level given by (2.129) and (2.130), i.e.,

$$E(\theta_k) = \alpha + 2\beta \cos \theta_k \tag{2.141}$$

[cf. (2.96), remembering that $E_0^s + \alpha_s \to \alpha$ in (2.141)]. Since $-1 \leq \cos \theta_k \leq 1$, the N-levels are distributed through a band of width $|4\beta|$, which lies between the band edges (Fig. 2.9)

$$E_{max} = \alpha - 2\beta \ ; \ \ E_{min} = \alpha + 2\beta, \tag{2.142}$$

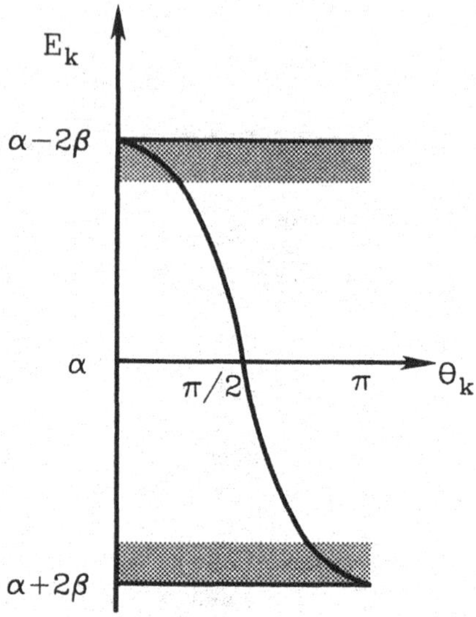

Fig. 2.9. Variation of E_k with θ_k across the band of width $|4\beta|$.

if β is assumed to be negative (cf. 2.97 and 2.98).

2.4 Nearly-Free Electron Approximation

In the TBA, the potential energy of an electron is assumed to account for nearly all the total energy, and the allowed energy bands are narrow compared with the FEG. The NFE approximation proceeds from the opposite point of view, namely, that the potential energy of an electron is small compared to its total energy. Contrary to the TBA, in the NFE model, the atoms of the crystal are assumed to be so close together that the wave functions of the electrons on neighbouring atoms overlap to a large extent. Under these conditions, the allowed bands will be broad and the FEG regions quite narrow.

The motions of electrons in a periodic field $V(\mathbf{r})$ are described by the Schrödinger equation

$$[-\boldsymbol{\nabla}^2 + V(\mathbf{r})]\psi_{\mathbf{k}}(\mathbf{r}) = E_{\mathbf{k}}\psi_{\mathbf{k}}(\mathbf{r}). \tag{2.143}$$

Substituting (2.11) in (2.143), and dividing through by $e^{i\mathbf{k}\cdot\mathbf{r}}$, leads to

$$-\boldsymbol{\nabla}^2 u_{\mathbf{k}} + V(\mathbf{r})u_{\mathbf{k}} - 2i(\mathbf{k}\cdot\boldsymbol{\nabla} u_{\mathbf{k}}) = (E_{\mathbf{k}} - \mathbf{k}^2)u_{\mathbf{k}}(\mathbf{r}). \tag{2.144}$$

Since $V(\mathbf{r})$ and $u_{\mathbf{k}}(\mathbf{r})$ are periodic, they can be expressed as a Fourier series of the form

$$V(\mathbf{r}) = \sum_{\mathbf{g}} V_{\mathbf{g}} e^{2\pi i(\mathbf{b_g}\cdot\mathbf{r})} \tag{2.145}$$

$$u_{\mathbf{k}}(\mathbf{r}) = \sum_{\mathbf{h}} a_{\mathbf{h}} e^{2\pi i(\mathbf{b_h}\cdot\mathbf{r})} \tag{2.146}$$

where $\mathbf{b_g}$ and $\mathbf{b_h}$ are the reciprocal lattice vectors (cf. 2.29). Without any loss of generality, the first term in (2.145) can be set equal to zero, i.e.,

$$V_{\mathbf{0}} = 0. \tag{2.147}$$

Thus, (2.145) to (2.147) in (2.144) lead to

$$\sum_{\mathbf{h}} (2\pi\mathbf{b_h})^2 a_{\mathbf{h}} e^{2\pi i(\mathbf{b_h}\cdot\mathbf{r})} + 2\sum_{\mathbf{h}} (2\pi\mathbf{b_h}\cdot\mathbf{k}) a_{\mathbf{h}} e^{2\pi i(\mathbf{b_h}\cdot\mathbf{r})}$$

$$+\sum_{\mathbf{h}}\sum_{\mathbf{g}\neq\mathbf{0}} V_{\mathbf{g}} a_{\mathbf{h}} e^{2\pi i[(\mathbf{b_g}+\mathbf{b_h})\cdot\mathbf{r}]} = \sum_{\mathbf{h}} (E_{\mathbf{k}} - \mathbf{k}^2) a_{\mathbf{h}} e^{2\pi i(\mathbf{b_h}\cdot\mathbf{r})}. \tag{2.148}$$

With the components of the vectors $\mathbf{g} = (g_1, g_2, g_3)$ and $\mathbf{h} = (h_1, h_2, h_3)$ being integers, $\mathbf{h}$ may be replaced by $\mathbf{h} - \mathbf{g}$, so that $\mathbf{b_g} + \mathbf{b_h} = \mathbf{b_h}$. The double sum in (2.148) then takes the form

$$\sum_{\mathbf{h}}\sum_{\mathbf{g}\neq\mathbf{0}} V_{\mathbf{g}} a_{\mathbf{h}-\mathbf{g}} e^{2\pi i \mathbf{b_h}\cdot\mathbf{r}} \tag{2.149}$$

and (2.148) can be written as

$$\sum_{\mathbf{h}} e^{2\pi i(\mathbf{b_h}\cdot\mathbf{r})} \left\{ a_{\mathbf{h}}[E_{\mathbf{k}} - (\mathbf{k} + 2\pi\mathbf{b_h})^2] - \sum_{\mathbf{g}\neq\mathbf{0}} V_{\mathbf{g}} a_{\mathbf{h}-\mathbf{g}} \right\} = 0. \tag{2.150}$$

The left-hand side of (2.150) will be identically equal to zero for all $\mathbf{r}$, if

$$a_{\mathbf{h}}[E_{\mathbf{k}} - (\mathbf{k} + 2\pi\mathbf{b_h})^2] - \sum_{\mathbf{g}\neq\mathbf{0}} V_{\mathbf{g}} a_{\mathbf{h}-\mathbf{g}} = 0. \tag{2.151}$$

Since $h_1,\ h_2,\ h_3 = 0,\ \pm1,\ \pm2, \ldots$, equation (2.151) represents an infinite linear homogeneous set of equations for the coefficients $a_{\mathbf{h}}$, whose non-zero values require the determinantal compatibility condition to be satisified. This leads to an infinite-order secular equation, whose infinite number of roots gives the *exact* values of the allowed energies $E_{\mathbf{k}}$ of the system, since no approximations have been made so far. Unfortunately, however, finding the energy values for the general case is extremely complicated in practice, so (2.151) can only be solved by making some simplifying assumptions.

(i) *Perfectly-free electrons*

For free electrons, $V(\mathbf{r}) = 0$, so all the coefficients $V_{\mathbf{g}} = 0$ in (2.145). Consequently, (2.151) becomes

$$a_{\mathbf{h}}[E_{\mathbf{k}} - (\mathbf{k} + 2\pi\mathbf{b_h})^2] = 0 \tag{2.152}$$

which is satisfied if, either

$$a_{\mathbf{h}} = 0 \tag{2.153}$$

or

$$E_{\mathbf{k}} - (\mathbf{k} + 2\pi\mathbf{b_h})^2 = 0. \tag{2.154}$$

However, for free electrons, the Bloch function (2.11) becomes the free-electron wave function

$$\psi_{\mathbf{k}}(\mathbf{r}) = Ce^{i\mathbf{k}\cdot\mathbf{r}} \tag{2.155}$$

so the function $u_{\mathbf{k}}(\mathbf{r}) = C$ (constant). From (2.146), it is clear that this can only happen if $\mathbf{b_h} = \mathbf{0}$, which implies that $\mathbf{h} = \mathbf{0}$, so (2.146) reduces to $u_{\mathbf{k}}(\mathbf{r}) = a_{\mathbf{0}}$ and the wave function (2.11) takes the form

$$\psi_{\mathbf{k}}(\mathbf{r}) = e^{i\mathbf{k}\cdot\mathbf{r}} \tag{2.156}$$

where, for convenience,

$$a_{\mathbf{0}} = 1. \tag{2.157}$$

Since $a_{\mathbf{0}} \neq 0$, equation (2.152) can only be satisfied for $\mathbf{h} = \mathbf{0}$ if (2.154) holds, which leads to the free-electron energy expression

$$E_{\mathbf{k}} = E_0(\mathbf{k}) = \mathbf{k}^2 \tag{2.158}$$

since $\mathbf{b_0} = \mathbf{0}$ by (2.29).

(ii) *Weak periodic field*

In this case, the electrons are *nearly* free, because the Fourier coefficients $V_{\mathbf{g}}$ in (2.145) are small, and may be treated as first-order corrections to the free-electron situation. Similarly, the function $u_{\mathbf{k}}(\mathbf{r})$ is *almost* independent of $\mathbf{r}$, so all the coefficients $a_{\mathbf{h}}$ in (2.146) are small compared with

a_0. It is therefore reasonable to approximate (2.151) by neglecting small quantities of second order, i.e., those involving the product $V_{\mathbf{g}}a_{\mathbf{h}-\mathbf{g}}(\mathbf{g} \neq \mathbf{h})$. Hence, (2.151) reduces to the single equation for which $\mathbf{g} = \mathbf{h}$, viz.,

$$a_{\mathbf{h}}[E_{\mathbf{k}} - (\mathbf{k} + 2\pi\mathbf{b_h})^2] - V_{\mathbf{h}}a_{\mathbf{0}} = 0 \tag{2.159}$$

which by (2.157) and (2.158) gives

$$a_{\mathbf{h}} = -\frac{1}{4\pi}\left[\frac{V_{\mathbf{h}}}{\pi\mathbf{b}_{\mathbf{h}}^2 + (\mathbf{b_h}\cdot\mathbf{k})}\right] \quad ,\mathbf{h} \neq \mathbf{0} \tag{2.160}$$

so the coefficients $a_{\mathbf{h}} \sim V_{\mathbf{h}}$, and represent a first-order correction to the wave function.

If the energy correction E' is defined as

$$E' = E_{\mathbf{k}} - E_0(\mathbf{k}) \tag{2.161}$$

then by (2.158)

$$E_{\mathbf{k}} = \mathbf{k}^2 + E'. \tag{2.162}$$

Substituting (2.162) and (2.160) into (2.151) yields

$$E'a_{\mathbf{h}} = -V_{\mathbf{h}} + \sum_{\mathbf{g}\neq\mathbf{0}} V_{\mathbf{g}}a_{\mathbf{h}-\mathbf{g}}. \tag{2.163}$$

With the aid of (2.147) and (2.157), setting $\mathbf{h} = \mathbf{0}$ in (2.163) gives

$$E' = \sum_{\mathbf{g}\neq\mathbf{0}} V_{\mathbf{g}}a_{-\mathbf{g}}. \tag{2.164}$$

Replacing $-\mathbf{g}$ by $\mathbf{g}$, and noting that $V_{-\mathbf{g}} = V_{\mathbf{g}}^*$ for $V(\mathbf{r})$ real, (2.164) becomes

$$E' = \sum_{\mathbf{g}\neq\mathbf{0}} V_{\mathbf{g}}^* a_{\mathbf{g}} \tag{2.165}$$

which by (2.160) leads to the second-order energy correction

$$E' = -\frac{1}{4\pi}\sum_{\mathbf{g}\neq\mathbf{0}} \frac{|V_{\mathbf{g}}|^2}{\pi\mathbf{b}_{\mathbf{g}}^2 + (\mathbf{b_g}\cdot\mathbf{k})}. \tag{2.166}$$

The corrections (2.160) and (2.166) are satisfactory, provided $(\pi\mathbf{b}_{\mathbf{g}}^2 + \mathbf{b_g}\cdot\mathbf{k})$ does not approach zero. If

$$\pi\mathbf{b}_{\mathbf{g}}^2 + \mathbf{b_g}\cdot\mathbf{k} = \mathbf{b_g}\cdot(\pi\mathbf{b_g} + \mathbf{k}) \simeq 0 \tag{2.167}$$

then neither the energy correction E' nor the corresponding coefficient $a_{\mathbf{h}}$ is small, and (2.160) and (2.166) cannot be used. In other words, for a wave vector $\mathbf{k}$ satisfying (2.167) for a certain vector $\mathbf{b_g}$, the electrons cannot be treated as being nearly free, because their motion is strongly perturbed. However, it is easy to see that (2.167) corresponds exactly to a Bragg reflection (Levy 1968) of an electron of wave vector $\mathbf{k}$ into the direction

$$\mathbf{k}' = \mathbf{k} + 2\pi\mathbf{b_g}. \tag{2.168}$$

Indeed, since the energies of the incident and reflected beams must be equal, $|\mathbf{k}'| = |\mathbf{k}|$, and (2.168) gives

$$\mathbf{k}'^2 = \mathbf{k}^2 = \mathbf{k}^2 + 4\pi^2\mathbf{b_g}^2 + 4\pi\mathbf{b_g}\cdot\mathbf{k} \tag{2.169}$$

which reduces to (2.167). Thus, when (2.167) is satisfied, the electron wave suffers a Bragg reflection from the atomic plane perpendicular to the vector $\mathbf{b_g}$.

Consider now the situation where an electron, with wave vector $\mathbf{k}$, satisfies (2.167) for a fixed vector $\mathbf{b_g}$, i.e., for a fixed value of $\mathbf{g}$. If the amplitude $V_{\mathbf{h}}$ in (2.160), for $\mathbf{h} = \mathbf{g}$, is not too close to zero, the coefficient $a_{\mathbf{g}}$ (and, by the same argument, $a_{\mathbf{0}}$ for $\mathbf{h} = \mathbf{0}$) is large and (2.151) reduces to two equations, namely, those for $\mathbf{h} = \mathbf{g}$ and $\mathbf{h} = \mathbf{0}$ (Peierls 1930, Wilson 1958), i.e.,

$$a_{\mathbf{g}}[E_{\mathbf{k}} - (\mathbf{k} + 2\pi\mathbf{b_g})^2] - V_{\mathbf{g}}a_{\mathbf{0}} = 0 \tag{2.170a}$$

$$a_{\mathbf{0}}(E_{\mathbf{k}} - \mathbf{k}^2) - V_{\mathbf{g}}a_{-\mathbf{g}} = 0. \tag{2.170b}$$

Replacing $\mathbf{g}$ by $-\mathbf{g}$, (2.170b) becomes

$$a_{\mathbf{0}}(E_{\mathbf{k}} - \mathbf{k}^2) - V_{-\mathbf{g}}a_{\mathbf{g}} = 0. \tag{2.171}$$

Equations (2.170a) and (2.171) are a pair of simultaneous equations for $a_{\mathbf{0}}$ and $a_{\mathbf{g}}$, whose non-trivial solutions require the determinantal compatibility condition to be fulfilled, which gives

$$[E_{\mathbf{k}} - (\mathbf{k} + 2\pi\mathbf{b_g})^2](E_{\mathbf{k}} - \mathbf{k}^2) - |V_{\mathbf{g}}|^2 = 0. \tag{2.172}$$

When (2.167) is satisfied, i.e., when

$$\mathbf{b_g}\cdot\mathbf{k} = -\pi\mathbf{b_g}^2 \quad \text{or} \quad \mathbf{k}^2 = (\mathbf{k} + 2\pi\mathbf{b_g})^2 \tag{2.173}$$

from (2.169), equation (2.172) yields

$$E_{\mathbf{k}} = \mathbf{k}^2 \pm |V_{\mathbf{g}}|. \tag{2.174}$$

Thus, for a wave vector $\mathbf{k}$ satisfying (2.173), a Bragg reflection occurs, which is accompanied by an energy discontinuity of $2|V_{\mathbf{g}}|$.

Returning to (2.170a), and writing

$$a_{\mathbf{g}} = V_{\mathbf{g}} a_0 [E_{\mathbf{k}} - (\mathbf{k} + 2\pi \mathbf{b}_{\mathbf{g}})^2]^{-1} \tag{2.175}$$

means that (2.146) now gives

$$u_{\mathbf{k}}(\mathbf{r}) = a_0 \left\{ 1 + V_{\mathbf{g}} e^{2\pi i \mathbf{b}_{\mathbf{g}} \cdot \mathbf{r}} / [E_{\mathbf{k}} - (\mathbf{k} + 2\pi \mathbf{b}_{\mathbf{g}})^2] \right\}$$

so the Bloch function (2.11) becomes

$$\psi_{\mathbf{k}}(\mathbf{r}) = a_0 \left\{ e^{i\mathbf{k}\cdot\mathbf{r}} + V_{\mathbf{g}} e^{i(\mathbf{k}+2\pi\mathbf{b}_{\mathbf{g}})\cdot\mathbf{r}} / [E_{\mathbf{k}} - (\mathbf{k} + 2\pi \mathbf{b}_{\mathbf{g}})^2] \right\}. \tag{2.176}$$

However, since

$$\pi \mathbf{b}_{\mathbf{g}} = -\mathbf{k} \tag{2.177}$$

from (2.173), it follows that (2.176) reduces to

$$\psi_{\mathbf{k}}(\mathbf{r}) = a_0 [e^{i\mathbf{k}\cdot\mathbf{r}} + V_{\mathbf{g}} e^{-i\mathbf{k}\cdot\mathbf{r}} / (E_{\mathbf{k}} - \mathbf{k}^2)] \tag{2.178}$$

or

$$\psi_{\mathbf{k}}(\mathbf{r}) = a_0 [e^{i\mathbf{k}\cdot\mathbf{r}} \pm (V_{\mathbf{g}}/|V_{\mathbf{g}}|) e^{-i\mathbf{k}\cdot\mathbf{r}}] \tag{2.179}$$

by (2.174).

For a one-dimensional system, where

$$\mathbf{b}_{\mathbf{g}} \equiv b_g = \pm g/a \ ; \quad g = 1, 2, 3, ... \tag{2.180}$$

(cf. 2.28 and 2.29), equation (2.173) gives the wave number

$$k = \mp \pi g / a \tag{2.181}$$

and (2.172) becomes

$$[E_k - (k + 2\pi b_g)^2](E_k - k^2) = |V_g|^2. \tag{2.182}$$

Figure 2.10(a) shows the graph of this equation, together with the free-electron curve $E_0(k) = k^2$. Comparing these two curves shows that the periodic potential has opened up FEG in the continuous free-electron curve, thus forming a series of allowed bands separated by FEG. The energy levels in the first band correspond to the wave numbers k in the first Brillouin zone, those in the second band to wave numbers in the second zone, and so on. This arrangement is referred to as the *extended zone scheme.*

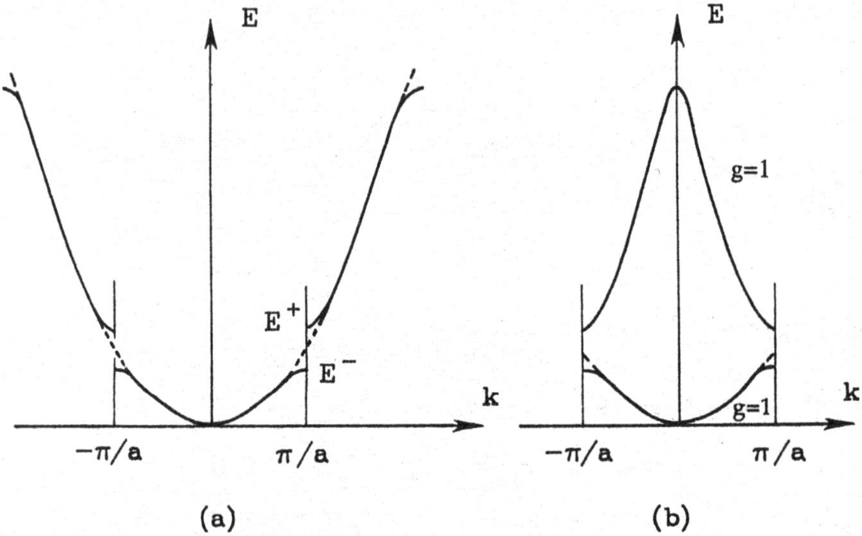

Fig. 2.10. Dependence of electronic energy E on wave number k for NFE approximation: (a) extended and (b) reduced zone scheme.

Differentiation of (2.182) with respect to k shows that extrema occur for k-values given by (2.181), for which

$$d^2E_k/dk^2 = 2[E^{\pm} + (\pi g/a)^2]/[E^{\pm} - (\pi g/a)^2] \tag{2.183}$$

and

$$E^{\pm} = (\pi g/a)^2 \pm |V_g| \tag{2.184}$$

by (2.174). Thus, a maximum (minimum) is located at $E^-(E^+)$ for each value of g (Fig. 2.10a).

If the k-values are restricted to the first Brillouin zone (i.e., $g = 1$)

$$-\pi/a \leq k \leq \pi/a \tag{2.185}$$

and E_k becomes a *multi-valued* function of k, as seen in Fig. 2.10(b). Such a procedure, called the *reduced zone scheme*, enables *all* the energy bands to be located in the first Brillouin zone region.

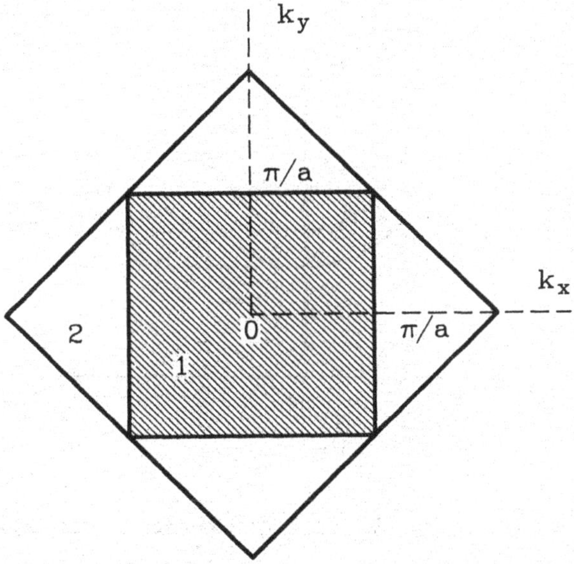

Fig. 2.11. First two Brillouin zones for simple square lattice.

The electron energy spectrum of a one-dimensional crystal is characterized by the fact that the successive bands of allowed energies are *always* separated by FEG. However, in two- and three-dimensional crystals, the neighbouring allowed bands may overlap each other, depending on the type of periodic potential. To illustrate this point, consider an NFE moving in the region of a simple square lattice away from the zone boundary with energy $E_k = (k_x^2 + k_y^2)$. The dependence of E_k on k_x and k_y is thus represented by a paraboloid of revolution about the E_k-axis. Sections of this paraboloid near the plane $k_y = \pm\pi/a$, corresponding to the horizontal edge of the first Brillouin zone (Fig. 2.11), are shown in Fig. 2.12. If the discontinuity in the energy $2|V_1|$, i.e., the distance between the parallel curves in Fig. 2.12, is sufficiently small, the upper limit of the lower band for $k_x = \pm\pi/a$ can lie *above* the lower limit of the upper band for $k_x = 0$, so that the allowed energy bands overlap.

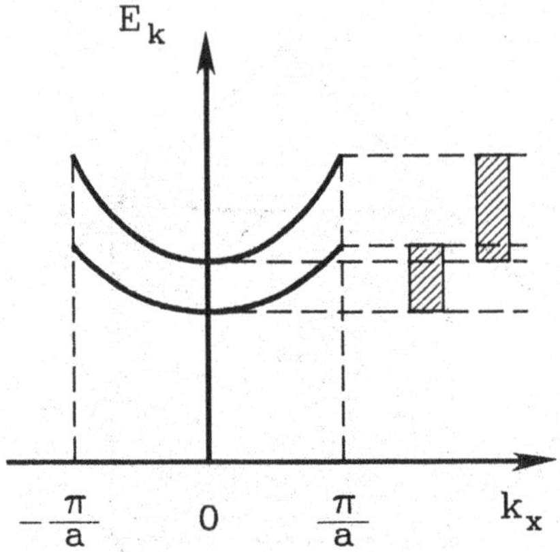

Fig. 2.12. Overlapping of energy bands for a simple square lattice.

At the first Brillouin zone boundary of a one-dimensional crystal, where $k = \pi/a$, (2.184) and (2.179) yield

$$E^{\pm} = (\pi/a)^2 \pm |V_1| \tag{2.186}$$

and

$$\psi^{\pm} = e^{i\pi x/a} \pm (V_1/|V_1|)e^{-i\pi x/a} \tag{2.187}$$

a_0 being set equal to unity. Accordingly, depending on the sign of V_1, two cases arise, viz.,

E	ψ	
	$V_1 > 0$	$V_1 < 0$
$(\pi/a)^2 + \mid V_1 \mid$	$\cos(\pi x/a)$	$\sin(\pi x/a)$
$(\pi/a)^2 - \mid V_1 \mid$	$\sin(\pi x/a)$	$\cos(\pi x/a)$

(2.188)

Hence, although the energies $E^{\pm}$ at the upper and lower edges of the first FEG are independent of the sign of V_1, the sign does play a role in the spatial symmetries of zone boundary wave functions, which are *standing waves* (Raimes 1967, Mark 1970).

Indeed, for an *attractive* potential ($V_1 < 0$) the charge densities are

$$\begin{aligned} \rho^+ &= |\psi^+|^2 \sim \sin^2(\pi x/a) \\ \rho^- &= |\psi^-|^2 \sim \cos^2(\pi x/a). \end{aligned} \tag{2.189}$$

Thus, if the ion cores are located at $x = 0, \pm a, \pm 2a, \ldots$, then the ψ^- functions accumulate charge (i.e. ρ^- maximum) *at* the ions, where it feels the most attraction and the energy E^- is lower. It also is worth noting that, since the maxima of the ψ^- functions correspond to the minima of the potential energy, any shift in the charge density curves to the left or right causes an increase in the energy. For the *repulsive* potential ($V_1 > 0$) case, where the ions repel the electrons, the reverse argument applies, namely, the charge densities are now

$$\begin{aligned} \rho^+ &= |\psi^+|^2 \sim \cos^2(\pi x/a) \\ \rho^- &= |\psi^-|^2 \sim \sin^2(\pi x/a) \end{aligned} \tag{2.190}$$

and the charge is concentrated *between* the ion cores at $x = \pm a/2, \pm 3a/2, \ldots$, for the minimum energy situation.

2.5 Mathieu Potential

One of the most obvious and simplest ways to represent a periodic potential in a one-dimensional crystal is by a sinusoidal wave. For such a potential, the Schrödinger equation reduces to the well-known *Mathieu equation* (1868), whose solutions have been thoroughly discussed by McLachlan (1951). The amplitude V_a of the sinusoidal wave is a measure of the crystal potential strength in a monatomic crystal, and the ionicity in a diatomic one. When V_a is small (large), the Mathieu potential model approaches the NFE (TBA) limit, and thus acts as a bridge between these two extreme descriptions.

The Mathieu potential received early recognition by Brillouin (1930) as an appropriate model for developing an energy band theory of solids, while Morse (1930) used it in his study of electron diffraction. Slater (1952) treated the Mathieu problem in one, two and three dimensions, in an attempt to describe the energy bands in a realistic crystal. More recently, Carver (1971) has discussed the symmetries of Mathieu functions, and the correspondence between the functions and the electron wave functions at the centres and edges of bands.

Consider a one-dimensional crystal with a sinusoidal potential of period a and maximum (repulsive) value V_a at $x = 0$, i.e., (Levine 1968, Davison and Levine 1970)

$$V(x) = V_a \cos(2\pi x/a). \tag{2.191}$$

The Schrödinger equation may now be written as

$$\psi''(z) + (\alpha - 2q\cos 2z)\psi(z) = 0 \tag{2.192}$$

where

$$\begin{aligned} z &= \pi x/a \ , \quad T = \pi^2/a^2 \\ \alpha &= E/T \ , \quad 2q = V_a/T. \end{aligned} \tag{2.193}$$

Solutions of the Mathieu equation (2.192) are best sought by first treating the free-electron case of $q = 0$, where the crystal potential is flat. The unnormalized standing wave solutions of (2.192) are then

$$\begin{aligned} &\psi(z) = \cos mz, \ \sin mz \\ &\alpha = m^2 \quad ; \quad m = 1, 2, \ldots \end{aligned} \tag{2.194}$$

For $q \neq 0$, both the energy parameter α and the wave function $\psi(z)$ must be functions of q, and may be expressed as power series, viz.,

$$\begin{aligned} \alpha(q) &= m^2 + \sum_{n=1}^{\infty} \alpha_n q^n \\ \psi(z, q) &= \cos\ mz + \sum_{n=1}^{\infty} c_n(z) q^n. \end{aligned} \tag{2.195}$$

Inserting (2.195) in (2.192) for $m = 1$, and equating coefficients of like powers of q gives

$$q^0 : \cos z - \cos z = 0 \tag{2.196}$$

$$q^1 : c_1'' + c_1 + (\alpha_1 - 1)\cos z = \cos 3z \tag{2.197}$$

$$q^2 : c_2'' + c_2 + \alpha_2 \cos z = 2c_1 \cos 2z - \alpha_1 c_1 \tag{2.198}$$

and so on. These equations are solved by using symmetry conditions and successive integrations.

For example, the particular solution of (2.197) is the non-periodic function $\frac{1}{2}(1-\alpha_1)z \sin z$, which must vanish if $\psi(z,q)$ is to be periodic, so

$$\alpha_1 = 1. \tag{2.199}$$

Thus, (2.197) becomes

$$c_1'' + c_1 = \cos 3z \tag{2.200}$$

whose particular solution is

$$c_1 = -(1/8) \cos 3z. \tag{2.201}$$

The complementary function $\sin z$ or $\cos z$ of (2.200) introduces a term of the form $q \sin z$ or $q \cos z$ into the solution (2.195). However, the coefficient of $\cos z$ in (2.195) is unity *for all* q, and $\sin z$ is odd, so the complementary functions of (2.200) and those of subsequent equations are omitted. Hence, (2.199) and (2.201) in (2.198) leads to

$$c_2'' + c_2 + (\tfrac{1}{8} + \alpha_2) \cos z = \tfrac{1}{8}(\cos 3z - \cos 5z). \tag{2.202}$$

Continuing with the above procedure, enables the energy and wave function coefficients α_n and c_n to be computed, so that $\alpha(q)$ and $\psi(q,z)$ in (2.195) may be found to any desired degree of accuracy.

Another solution of (2.192), viz.,

$$\psi(z,q) = \sin mz \; + \sum_{n=1}^{\infty} s_n(z) q^n \tag{2.203}$$

can also be generated in the same way. The standing wave eigenfunctions (2.195) and (2.203) so obtained are called the *cosine* and *sine elliptic functions* $ce_m(z,q)$ and $se_m(z,q)$, respectively. The corresponding eigenvalues $\alpha = a_m$ and $\alpha = b_m$ are known functions of q. Tables and graphs of m and q are available (Jahnke and Emde 1945, McLachlan 1951), and their most relevant properties are shown in Figs. 2.13 and 2.14.

In the NFE limit, as $q \to 0$, the lattice potential becomes flat and $ce_m(z,q) \to \cos mz$, while $se_m(z,q) \to \sin mz$. These are the free-electron wave functions represented by the broken curves in Fig. 2.13. As q increases, the attractive potential well at $z = \pi/2$ (and $3\pi/2$) deepens, causing the eigenfunction amplitudes and nodes to move closer to this point to take advantage of the attractive potential found there. For very large q, the eigenfunctions become more like LCAO functions (solid curves in Fig. 2.13), and resemble s, p, d, etc., orbitals. In particular, the $ce_0(z, q)$ and $se_1(z, q)$ functions become s-like (no nodes near $z = \pi/2$), $ce_1(z, q)$ and $se_2(z, q)$ become p-like (one node at $z = \pi/2$), $ce_2(z, q)$ and $se_3(z, q)$ become d-like (two nodes near $z = \pi/2$), etc. Thus, Mathieu functions display a smooth mathematical and conceptual transition between the NFE and TBA limits.

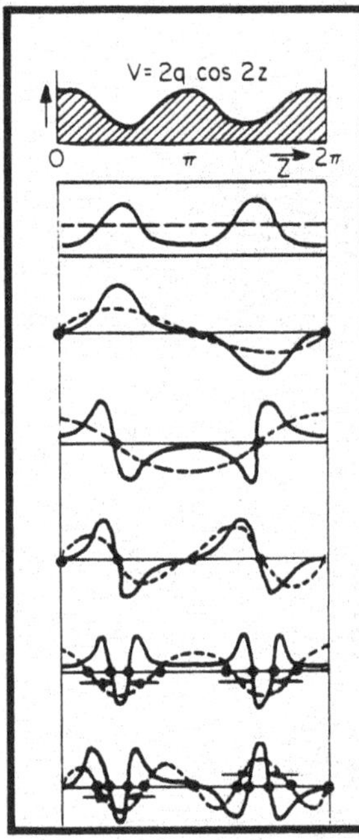

Function	NFE Limit q → 0	LCAO Limit q >> 0
$ce_0(z,q)$	1	s(+)
$se_1(z,q)$	sin z	s(−)
$ce_1(z,q)$	cos z	p(+)
$se_2(z,q)$	sin 2z	p(−)
$ce_2(z,q)$	cos 2z	d(+)
$se_3(z,q)$	sin 3z	d(−)

Fig. 2.13. Some properties of Mathieu functions. (+) denotes bonding and (−) antibonding orbitals. (After Levine 1968.)

The band structure diagram is shown in Fig. 2.14(a), where the $\alpha(= a_m$ or $b_m)$ versus q curves are drawn. Here, the allowed Bloch-like states $e^{\pm ikz}\phi(\pm z, q)$ occur in the shaded bands, while in the FEG's between the bands the wave functions have the forms $e^{\mp\mu z}\phi(\pm z, q)$. The nth $(m = n)$

gap is bounded by the (a_n, b_n) states, a_0 being the state at the bottom of the first band. Notice that there are no gaps for $q = 0$, the free-electron limit, and there is no band crossing. As q increases, the allowed bands become narrower (more atomic-like), and the narrow-band energies are pulled down lower to take advantage of the greater attractive potential near $z = \pi/2$.

Curves very similar to Fig. 2.14(a) have been obtained by Van der Pol and Strutt (1928) for the rectangular-type of KP potential (Fig. 2.1). A description of their analysis has been provided by Brillouin (1953), who also discussed the Mathieu problem.

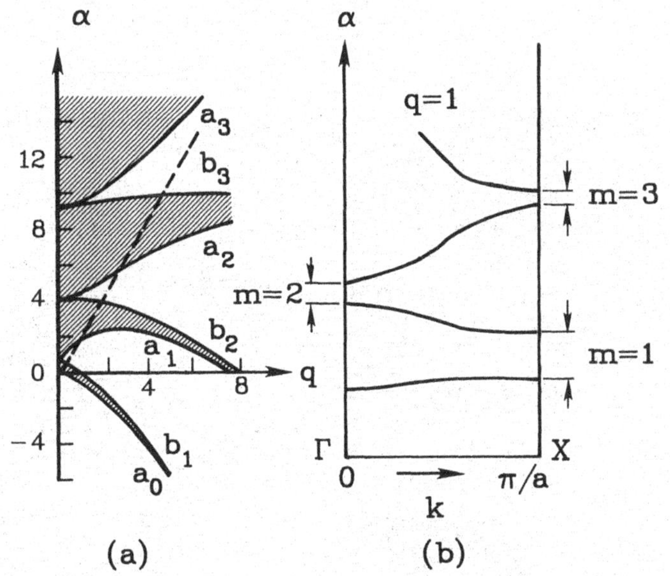

Fig. 2.14. Variation of energy bands (a) with potential amplitude q and (b) with wave number k for $q = 1$. (After Levine 1968.)

Fig. 2.14(b) shows the α vs. k Brillouin-zone representation of Fig. 2.14(a) for $q = 1$. The first ($m = 1$) FEG occurs at the edge X of the Brillouin zone, the second ($m = 2$) at the centre Γ, and so on. Thus, FEG at X and Γ correspond to m odd and m even, respectively.

3

ELECTRONS AT CRYSTAL SURFACES

God made solids, but surfaces were the work of the Devil.
— Wolfgang Pauli

Since crystals in their natural state are *finite*, i.e., limited by surfaces, the question arises as to how the energy spectrum of an electron propagating in a finite crystal differs from that in an infinite one. The partial answer is provided by using the Born-von Kármán (1912, 1913) boundary condition (2.138), which shows that, if a finite crystal of N atoms is cut from an infinite one, then the previous continuous *allowed* band becomes a sequence of N *discrete* levels (2.141). Clearly, this is not a big difference, if the crystal (i.e., N) is very large. However, there is another important consequence of taking the crystal size into account, namely, that the presence of the surface produces discrete energy levels, which lie in the FEG regions that are inaccessible to electrons in infinite crystals. Tamm (1932) was the first to demonstrate the existence of these *surface states* by terminating the KP potential as a means of representing the surface. Because of the importance of Tamm's work in establishing the new idea of surface states, we shall now briefly discuss his findings.

3.1 Concept of Surface States

According to the Bloch-Floquet theorem (2.3), the wave number k is determined by the boundary conditions. For an infinite crystal, k is taken to be *real*, to ensure that the wave function remains finite. However, when a crystal is limited by a surface, such a restriction is no longer necessary, and k may also assume *complex* values. A more detailed investigation of this problem can be performed by solving the Schrödinger equation for an electron moving in the potential field schematically shown in Fig. 3.1, where the potential is periodic inside the crystal ($x > 0$) and constant in the vacuum region ($x < 0$). Thus, the plane $x = 0$ represents the surface.

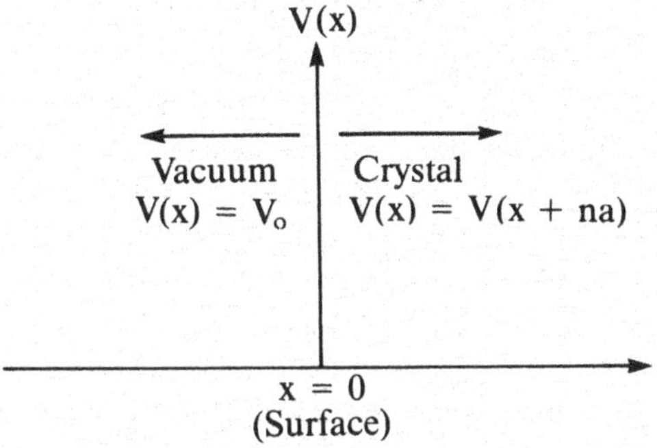

Fig. 3.1. Schematic diagram of crystal-vacuum interface.

In studying the energy spectrum of an electron in such a semi-infinite crystal lattice, the crux of Tamm's approach was to match the wave function and its derivative in the vacuum region to those inside the crystal (the so-called *direct matching procedure*). To preserve the finite nature of the wave function as $x \to -\infty$, the solution of the Schrödinger equation in the vacuum region ($x \leq 0$) has to be taken as

$$\psi_v = De^{k_0 x} \tag{3.1}$$

where

$$k_0 = (V_0 - E)^{1/2} \tag{3.2}$$

is positive and real, i.e., $E < V_0$. Inside the crystal ($x > 0$)

$$\psi_c(x) = Au_k(x)e^{ikx} + Bu_{-k}(x)e^{-ikx} \tag{3.3}$$

(cf. 2.10). Performing the matching at $x = 0$ leads to the following relations between the constants A, B and D:

$$\begin{aligned} Au_k(0) \;&+\; Bu_{-k}(0) = D \\ A[u_k'(0) \;&+\; iku_k(0)] + B[u_{-k}'(0) - iku_{-k}(0)] = Dk_0 \,. \end{aligned} \tag{3.4}$$

Consider now two cases:

(i) *k real*, i.e., the electron energy E lies inside one of the allowed energy bands of an infinite crystal. The wave function (3.3) is finite for any A and B, so the matching conditions (3.4) become two linear equations for three unknowns A, B and D. Obviously, such a set of equations always has solutions for any values of the coefficients, i.e., for any energy E inside the allowed band. Consequently, all the electronic energy levels, which are allowed in the infinite crystal, are also allowed in the semi-infinite one.

(ii) *k complex*, i.e., the corresponding energies lie inside the FEG of the infinite crystal. For a semi-infinite crystal ($x > 0$), the wave function (3.3) remains finite, when $x \to \infty$, only if A or B equals zero (depending on the sign of the imaginary part of k). Hence, (3.4) become two linear equations for two unknowns and only have solutions for the values of the electron energy E which make the determinant of the system vanish. Therefore, in the semi-infinite crystal, the discrete energy levels can exist only in that part of the energy spectrum which corresponds to the FEG in the infinite crystal. It follows from (3.1) that the wave functions describing these levels are monotonically damped in the direction of the vacuum. Inside the crystal, they are also damped, but in an oscillatory fashion. This can be easily seen from (3.3), remembering that k is complex and that the coefficient, for the term increasing with x, is equal to zero. Consequently, an electron, whose energy corresponds to one of the discrete levels is *localized* near the crystal surface and, for this reason, the energy levels located in the FEG are known as *surface states*.

3.2 Tamm's Model

As a simple example, enabling some quantitative discussion, Tamm (1932) chose the terminated KP model (Fig. 3.2) in which the potential is represented by a sum of δ-barriers[1] (§2.2). The corresponding dispersion relation (2.47) is

$$\cos ka = \cos\xi + p\xi^{-1}\sin\xi \tag{3.5}$$

where

$$\xi = aE^{1/2}. \tag{3.6}$$

[1]It should be noted that many papers exist, which treat the δ-*well* type of crystal potential, i.e., $V(x) = -\,p\,\sum_{n=0}^{\infty}\,\delta(x - na)$, where p and x are positive. Here, the standard matching procedure at the surface ($x = 0$) requires the boundary conditions $\psi_v(0) = \psi_c(0)$ and $\psi_c'(0^+) - \psi_v'(0^-) = -\,(p/a)\psi_c(0)$, which follows from the integration of the Schrödinger equation across the δ-well potential discontinuity at $x = 0$. A detailed discussion of the δ-well model has been provided by Stęślicka (1974).

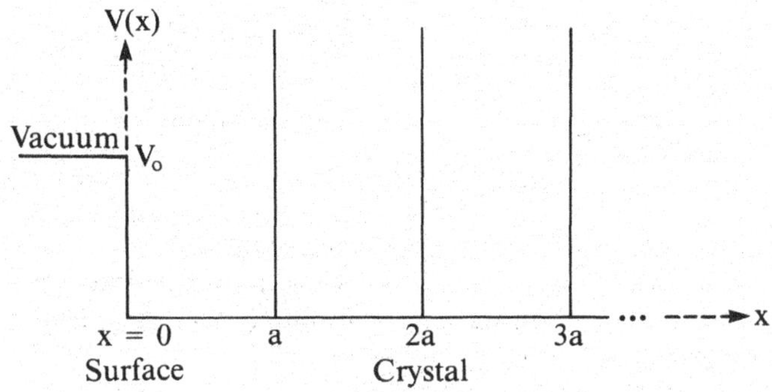

Fig. 3.2. Tamm's model of semi-infinite linear crystal with surface potential $V_0 > E$.

For any value of ξ, i.e., of the electron energy E, (3.5) has two roots k which differ only in sign and correspond to two particular solutions of the Schrödinger equation, i.e., to the first and second terms in (3.3). The complex form of k is taken as

$$k = i\mu + m\pi/a, \qquad m = 0, 1, 2, \ldots \tag{3.7}$$

so that

$$\cos ka = (-1)^m \cosh \mu a$$

where m identifies the $(m + 1)$ FEG, and μ is real, and may be either positive or negative. It is customary to assume

$$\mu > 0 \tag{3.8}$$

which, for $x > 0$, leads to $B = 0$ in (3.3).

The wave function of an electron inside the KP crystal, in the range $0 < x < a$, is in general, a linear combination of the solution given by (2.53) and (2.54) and the corresponding one for $-k$, i.e.,

$$\psi_c(x) = A(e^{ik_1x} + \lambda e^{-ik_1x}) + B(e^{ik_1x} + \lambda_- e^{-ik_1x}) \tag{3.9}$$

with

$$\lambda = (e^{ika} - e^{ik_1a})/(e^{-ik_1a} - e^{ika}) \tag{3.10}$$

and

$$\lambda_{-} = (e^{-ika} - e^{ik_1 a})/(e^{-ik_1 a} - e^{-ika}). \tag{3.11}$$

Using the Bloch-Floquet theorem (§2.1.1), immediately shows that, for the $(n+1)$th cell ($na < x < (n+1)a$, $n = 0, 1, 2 \ldots$), (3.9) takes the form

$$\begin{aligned} \psi_c(x) &= Ae^{ikna}[e^{ik_1(x-na)} + \lambda e^{-ik_1(x-na)}] \\ &+ Be^{-ikna}[e^{ik_1(x-na)} + \lambda_{-} e^{-ik_1(x-na)}]. \end{aligned} \tag{3.12}$$

For k complex, and of the form (3.7), the second term in (3.12) becomes infinitely large when $x \to \infty$, i.e., $n \to \infty$, so only the first term in (3.12) and (3.9) represents the appropriate solution of the Schrödinger equation. Consequently, for $0 < x < a$,

$$\psi_c(x) = A(e^{ik_1 x} + \lambda e^{-ik_1 x}) \tag{3.13a}$$

and

$$\lambda = [(-1)^m e^{-\mu a} - e^{ik_1 a}]/[e^{-ik_1 a} - (-1)^m e^{-\mu a}]; \quad m = 0, 1, 2, \ldots \tag{3.13b}$$

From (3.13), it follows that

$$\psi_c(a) = (-1)^m \psi_c(0) e^{-\mu a}$$

so, the wave function (3.13) decreases along the positive x-axis and can be used to study the localized electronic states.

Matching the wave functions (3.1) and (3.13) and their derivatives at $x = 0$, and solving the determinantal compatibility condition for k complex, (3.7), leads to

$$(-1)^m e^{-\mu a} = \cos\xi + (q^2 - \xi^2)^{1/2} \xi^{-1} \sin\xi \tag{3.14}$$

where

$$q = aV_0^{1/2}; \qquad q > \xi \tag{3.15}$$

and

$$m\pi < \xi < (m+1)\pi; \qquad m = 0, 1, 2, \ldots \tag{3.16}$$

On the other hand, for k complex, the dispersion relation (3.5) becomes

$$(-1)^m \cosh \mu a = \cos\xi + p\xi^{-1} \sin\xi \, . \tag{3.17}$$

Obviously, (3.14) and (3.17) have to be satisfied simultaneously, which leads to

$$\xi \cot\xi = (q^2/2p) - (q^2 - \xi^2)^{1/2} \, . \tag{3.18}$$

Equation (3.18) gives the dependence of the surface-state energies ($E = \xi^2/a^2$) on the parameters q and p, describing the surface and bulk properties, respectively (3.15 and 2.46), and is called the *surface-state energy expression.* It should be remembered that (3.18) was obtained under the assumption (3.8), i.e., $\mu > 0$. The equation defining μ can be obtained by subtracting (3.14) from (3.17) to give

$$(-1)^m \sinh \mu a = [p - (q^2 - \xi^2)^{1/2}]\xi^{-1} \sin \xi \; . \qquad (3.19)$$

Because of (3.16),

$$\mathrm{sgn}(\xi^{-1} \sin \xi) = (-1)^m$$

so $\mu > 0$ requires

$$p - (q^2 - \xi^2)^{1/2} > 0 \; . \qquad (3.20)$$

Since each root of (3.18) corresponds to a surface-state energy only if (3.20) is valid, this inequality is often called the *existence condition.* A graphical solution of (3.18) is presented in Fig. 3.3; in each interval $[m\pi, (m+1)\pi]$, (3.18) always has one root ξ, provided $m \geq 1$. For the first FEG ($m = 0$), (3.18) has a solution only if the inequality

$$q^2/2p - q < 1 \qquad (3.21)$$

is valid (Fig. 3.3). Thus, it may be concluded that in any FEG one surface state can exist provided (3.18) has roots which also satisfy (3.21).

In order to study the localization properties of an electron in a surface state, we employ the relative probability density function defined by (Thakkar and Stęślicka 1978)

$$P(x) = |\psi(x)|^2 \; / \; |\psi(0)|^2 \qquad (3.22)$$

where $\psi(x)$ is the surface-state wave function. Since the crystal surface is located at $x = 0$, $P(x)dx$ is the probability of finding an electron between x and $x + dx$ relative to the probability of finding it at the surface. Note that, using (3.22) instead of the standard $|\psi(x)|^2$ enables the normalization problem to be avoided, without any loss of generality.

The most direct way of obtaining the explicit form of $P(x)$ for Tamm's model is to use the wave function $\psi(x)$ given by (2.76), i.e., in the form derived by the Green-function technique. Using (2.76), (2.9) and (3.7) in (3.22) immediately lead to

$$P(x) = r^{2n} \left\{ r \sin \xi \left(\frac{x}{a} - n \right) - \sin \xi \left[\frac{x}{a} - (n+1) \right] \right\}^2 / \sin^2 \xi \qquad (3.23)$$

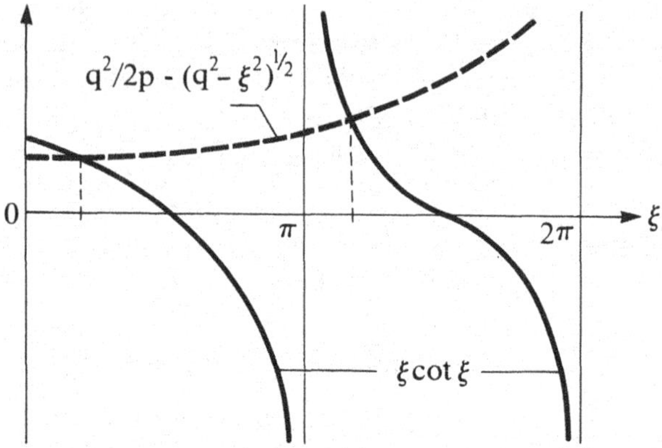

Fig. 3.3. Graphical solutions of Tamm's surface-state energy expression (3.18).

for $na < x < (n+1)a$ with $n = 0, 1, 2, 3, \ldots$ In (3.23),

$$r = (-1)^m e^{-\mu a} \tag{3.24}$$

while ξ, being a solution of (3.18) subject to (3.20), determines the energy E of the surface state (3.6). Outside the crystal, (3.1), (3.2), (3.6) and (3.15) in (3.22) lead to

$$P(x) = e^{(q^2-\xi^2)^{1/2}x/a}, \quad x < 0\,. \tag{3.25}$$

As an illustration of these simple analytical considerations, Fig. 3.4 shows $P(x)$ for a surface state in the first FEG. It is evident that electrons in surface states tend to be strongly localized inside the crystal near the surface. It is also interesting to note that, in the second or higher FEG, electrons do not exhibit pronounced localization (Fig. 3.5). In general, for surface states in the nth FEG, $P(x)$ has n nodes and n local maxima of equal amplitude in each unit cell. The amplitudes become progressively smaller as the unit cells are chosen further and further inside the crystal. In other words, $P(x)$ displays a damped oscillatory behaviour as x increases from zero at the surface to more positive values in the crystal.

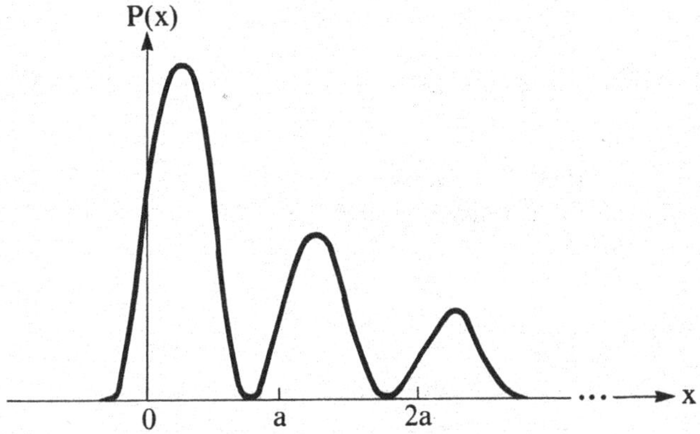

Fig. 3.4. Relative probability density function for electron in Tamm state in first FEG.

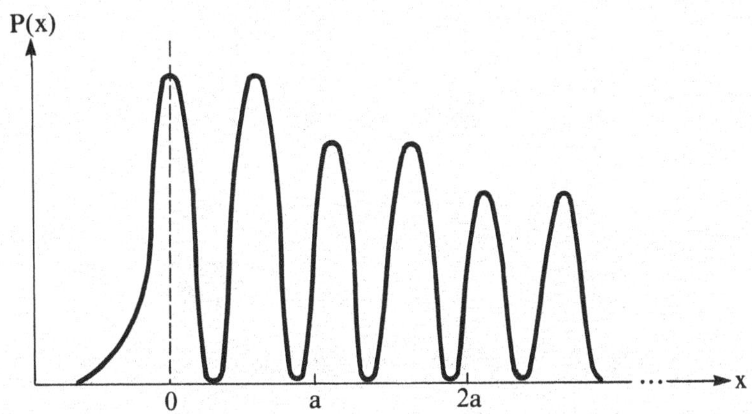

Fig. 3.5. $P(x)$ for second FEG.

It is noteworthy that a Tamm-like model was adopted recently for semiconductor superlattices and the existence of surface states shown explicitly (Stęślicka *et al.* 1990). Superlattices form, in fact, one-dimensional periodic structures, and the first experimental observation of 'surface states in the pure form that Tamm investigated theoretically' has been reported (Ohno *et al.* 1990). Thus, in superlattices, experimental reality seems to approach theoretical models and assumptions (cf. Esaki 1984, 1986).

3.3 Finite Crystal

An obvious extension of Tamm's work is to consider a finite crystal terminated by *two* surfaces. Although Sokolov (1934), and more recently Stęślicka (1974), have studied this problem, here a simpler and more direct approach is adopted.

3.3.1 Model

As shown in Fig. 3.6, a finite crystal can be represented by the δ-barrier KP model terminated on both sides, i.e., at $x = 0$ and $x = (N+1)a$, $(N+1)$ being the number of atoms (crystal cells), and a the lattice constant.

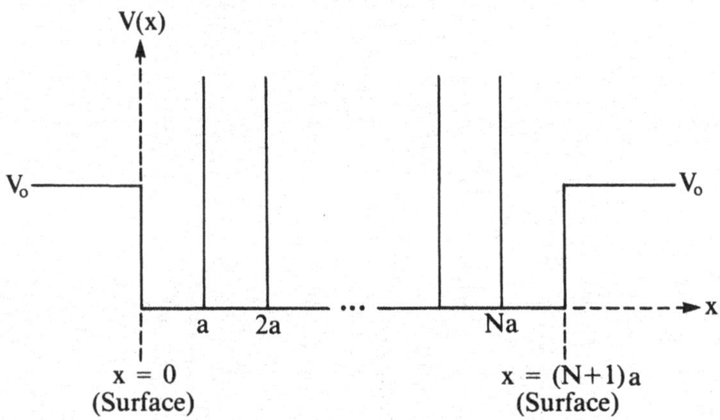

Fig. 3.6. Finite linear chain of δ-barriers.

The wave functions inside the crystal $(0 < x < (N+1)a,\ N = 0, 1, 2, \ldots)$ are given by (3.9) and (3.12) for the first and $(N+1)$th cells, respectively, while, in the vacuum regions, (3.1) represents the solution for $x < 0$ and

$$\psi_v(x) = C\ e^{-k_0[x-(N+1)a]}, \quad x > (N+1)a \tag{3.26}$$

where k_0 is given by (3.2). Note that, contrary to a semi-infinite case, the wave number k can now take real as well as complex values.

For the direct matching procedure, the boundary conditions are

$$\begin{aligned} \psi_v(0) = \psi_c(0)\ , &\qquad \psi_v[(N+1)a] = \psi_c[(N+1)a] \\ \psi_v'(0) = \psi_c'(0)\ , &\qquad \psi_v'[(N+1)a] = \psi_c'[(N+1)a]\ . \end{aligned} \tag{3.27}$$

Applying (3.27) to (3.1), (3.9), (3.12) and (3.26) leads to a set of four homogeneous equations for four unknowns. After some tedious algebra, expanding the resulting determinant gives the expression for the energy $E = \xi^2/a^2$ of an electron in a finite crystal, namely,

$$R^2 \sin(N+1)\theta - 2\xi R \sin N\theta + \xi^2 \sin(N-1)\theta = 0 \tag{3.28}$$

where

$$R = \xi \cos \xi + (q^2 - \xi^2)^{1/2} \sin \xi; \qquad q > \xi \tag{3.29}$$

and

$$\theta = ka \tag{3.30}$$

plays the role of the wave number, which can be real or complex.

Equation (3.28) has $(N+1)$ roots for $\theta (0 \leq \theta \leq \pi)$ corresponding to $(N+1)$ energy levels, as required for a crystal containing $(N+1)$ atoms (Fig. 3.6), and has the same form as (4.9) obtained by Goodwin (1939b) for a finite linear chain of atoms via the TBA. The following graphical analysis is based, to a large extent, on his treatment.

(i) *Graphical solution of (3.28)*

Solving (3.28) gives

$$\frac{\xi}{R} = \frac{\sin N\theta \pm \sin \theta}{\sin(N-1)\theta} \tag{3.31}$$

which leads to the two equations

$$\frac{\xi - R}{\xi + R} = \cot \tfrac{1}{2} N\theta \tan \tfrac{1}{2}\theta \equiv Y_1(\theta) \tag{3.32a}$$

$$\frac{\xi - R}{\xi + R} = -\tan \tfrac{1}{2} N\theta \tan \tfrac{1}{2}\theta \equiv Y_2(\theta)\ . \tag{3.32b}$$

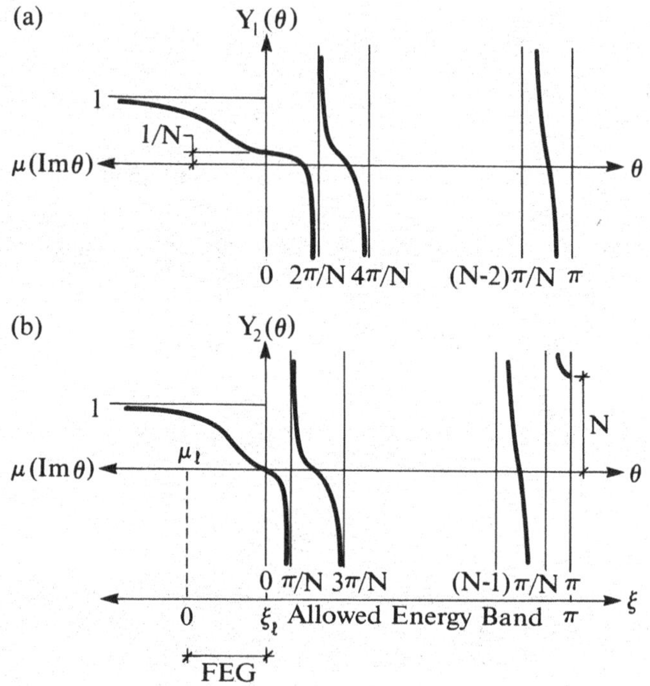

Fig. 3.7. Graphs of trigonometric functions (3.32) for θ real and pure imaginary ($\theta = i\mu a$).

A graph of $Y_1(\theta)$ and $Y_2(\theta)$ for θ real and imaginary ($\theta = i\mu a$) is shown in Fig. 3.7 for the case of even N (the case of N odd is similar). The correspondence between the values of θ and ξ, following from (2.47), is also displayed; ξ_ℓ in Fig. 3.7 is the upper limit of the first FEG obtained from the equation (cf. 2.47 and Fig. 2.3)

$$\cos \xi_\ell + p\xi_\ell^{-1} \sin \xi_\ell = 1$$

while the lower limit of the first FEG is

$$\mu_\ell a = \cosh^{-1}(1 + p)$$

as follows immediately from (2.47) for $\xi = 0$ and $\theta \equiv ka = i\mu a$.

It is easy to show that the limiting values of (3.32) are

$$\lim_{\substack{\xi\to\pi \\ \theta\to\pi}} \frac{\xi - R}{\xi + R} = \infty \tag{3.33}$$

$$\lim_{\substack{\xi\to 0 \\ \theta\to i\mu_\ell a}} \frac{\xi - R}{\xi + R} = -\frac{q}{q+2} < 0 \tag{3.34}$$

since $q > 0$ via (3.15). Therefore, both equations (3.32) together can have $(N + 1)$ real roots. However, when

$$(\xi - R)/(\xi + R) > 1/N \tag{3.35}$$

the lowest two roots disappear leaving only $(N - 1)$ real roots (Fig. 3.7). It is obvious that the two missing roots must become pure imaginary, since complex roots lead to complex values of energy, as follows from the KP relation (2.47), i.e.,

$$\theta = i\mu a. \tag{3.36}$$

Although imaginary values of the wave number give rise to energy levels in the FEG, the two discrete levels mentioned above become surface states only if electrons occupying them are localized near the crystal surface, as is discussed in (iii).

(ii) *Limiting case: N large*

When N is very large, it is reasonable to assume that

$$N \to \infty. \tag{3.37}$$

Using (3.36) and (3.37) reduces (3.32a) and (3.32b) to

$$\frac{\xi - R}{\xi + R} = \tanh \tfrac{1}{2}\, \mu a \tag{3.38}$$

which can be written as

$$e^{-\mu a} = \cos\xi + \xi^{-1}\sqrt{q^2 - \xi^2}\ \sin\xi \tag{3.39}$$

in accord with (3.14), which is not surprising, since the limit $N \to \infty$ changes the finite crystal model into a semi-infinite one. Equation (3.39) does *not* contain N, and describes only imaginary values of the wave number

μ corresponding to the surface state energies, viz., one doubly degenerate solution for $0 \leq \xi \leq \pi$, i.e., one surface state in the FEG (Fig. 3.3).

If N is large, but still finite, then physically it means that the effect of the surface at $x = (N+1)a$ on the phenomena occurring at $x = 0$ is negligible (and vice versa). Mathematically, this can be written as

$$\psi_c[x = (N+1)a] = 0 \ . \tag{3.40}$$

The boundary conditions (3.27) now reduce to $\psi_v(0) = \psi_c(0)$ and $\psi_v'(0) = \psi_c'(0)$, which, together with (3.40) and (3.12), lead to the energy expression

$$\sin(N+1)\theta / \sin N\theta = \xi / R \tag{3.41}$$

with R given by (3.29). Since an equation of this type arises in the context of the TBA (Chapter 4), a discussion of its solutions is given there. Note also, that (3.41) follows from (3.31), since the discriminant of (3.28) vanishes for $N \to \infty$.

However, it is worth noting here that, when $\theta = i\mu a$,

$$\lim_{N\to\infty} \sinh\,(N+1)\ \mu a / \sinh N\mu a = e^{\mu a}. \tag{3.42}$$

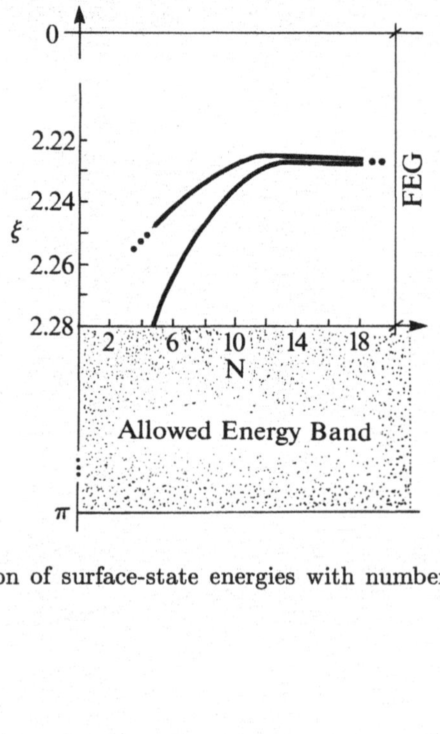

Fig. 3.8. Variation of surface-state energies with number of atoms in finite crystal.

Hence, in view of (3.29), equation (3.41) takes the form of (3.14) for a semi-infinite crystal.

(iii) *Numerical results*

To illustrate the analytical considerations, the energy expression (3.31), together with the KP relation (2.47), has been solved numerically for $\theta = i\mu a$, $\mu > 0$, fixed values of model parameters (q, p, a) and different values of N. The computed dependence of ξ on N is shown in Fig. 3.8, from which it is evident that the difference in the energies of two solutions $\Delta\xi$ is negligible, even for N small. For example, when $N = 10$, $\Delta\xi \sim 4 \times 10^{-8}$, which is equivalent to $\Delta E \sim 10^{-6}$eV. For $N > 20$, both solutions coalesce, so the use of the much simpler semi-infinite or quasi-finite (with 3.40) crystal models is completely justified.

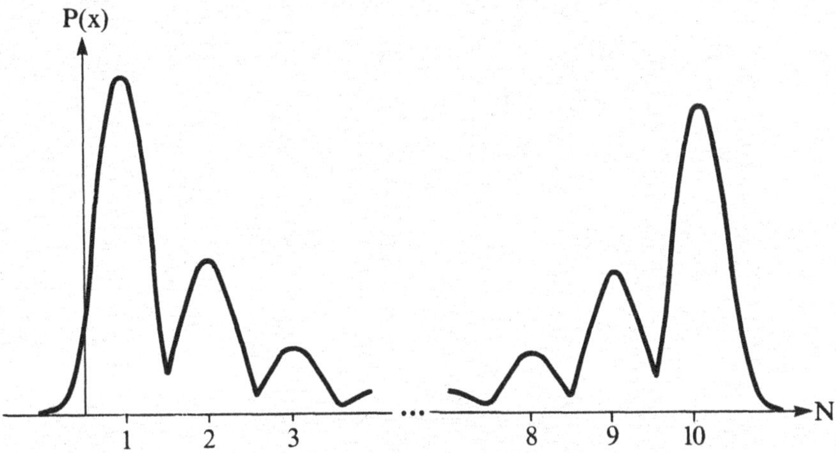

Fig. 3.9. $P(x)$ description of surface-state wave function in finite crystal of $N = 10$ atoms.

The localization properties of electrons in the two energy levels, obtained from the solution of (3.31), are presented in Figs. 3.9 and 3.10 for $N = 10$ and 20, respectively, with the aid of the relative probability density function $P(x)$, defined in (3.22). Since $P(x)$ is damped on both sides of the surfaces, electrons occupying these discrete levels are localized near the surfaces. Consequently, the two energy levels in the FEG can be recognized as *surface states.*

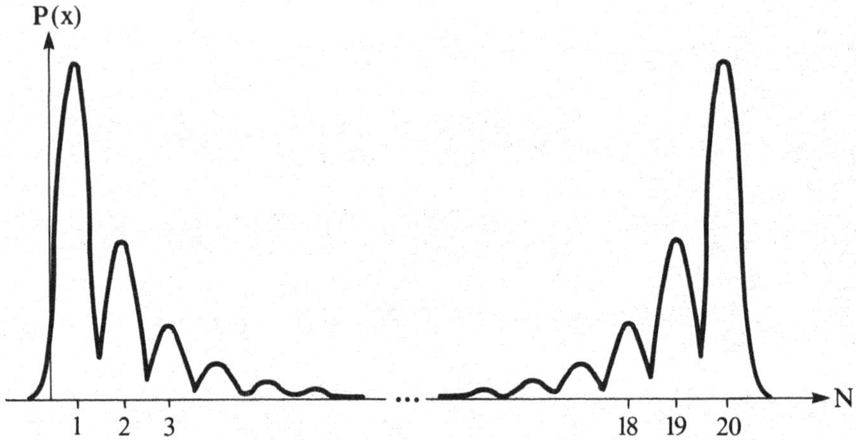

Fig. 3.10. Fig. 3.9 for $N = 20$.

4

TIGHT-BINDING APPROXIMATION

Having precise ideas often leads to a man doing nothing.
— Paul Valéry

In the previous chapter, surface states were analysed by means of a potential approach, which is a method that has been developed mainly by physicists. However, the *tight-binding approximation* (TBA) is based on the electron-orbital concept (§2.3), which, while commonly used in solid-state theory, has its origins in quantum chemistry.

4.1 Tamm States of a Monatomic Chain

The *bulk* states of a monatomic chain have already been described in the TBA treatment of the cyclic crystal (§2.3.2). If now, the bond is broken between the first and Nth atoms of a cyclic chain, then a linear chain of N atoms numbered $n = 1, 2, \ldots, N$ is formed. The bond-breaking process perturbs the electronic environment of the end atoms at $n = 1$ and N. Thus, we have a *finite* monatomic chain, where all the atoms, except those at $n = 1$ and N, are characterized by their Coulomb integrals α and the bonds between them by their resonance integrals β. If a p-orbital is associated with each atom, then $\beta > 0$ (2.107 and 2.116). The perturbations, introduced at the $n = 1$ and N atoms by creating the surfaces, are taken into account by changing α to α' at these atoms (Fig. 4.1).
Goodwin (1939b) was the first to study the electronic states of such a system within the framework of the TBA. Mathematically, these states are described by the solutions to the finite difference equation (2.125), viz.,

$$(E - \alpha)c_n = \beta(c_{n+1} + c_{n-1}), \quad n \neq 1 \text{ or } N \tag{4.1}$$

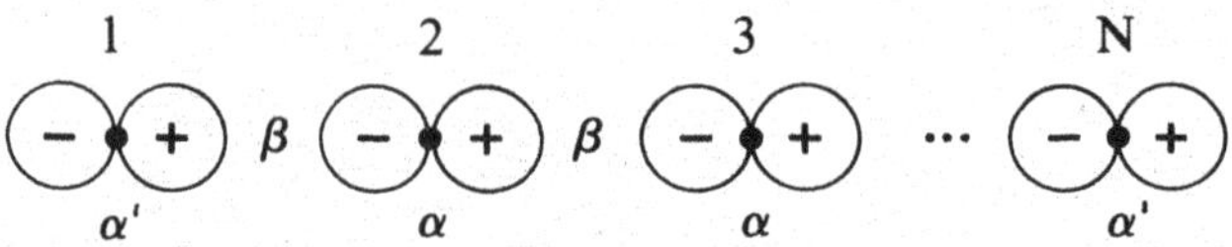

Fig. 4.1. Monatomic chain of p-orbitals with surfaces at $n = 1$ and N.

subject to the boundary conditions (Goodwin 1939b)

$$(E - \alpha')c_1 = \beta c_2 \tag{4.2}$$

$$(E - \alpha')c_N = \beta c_{N-1}. \tag{4.3}$$

Following the analysis of §2.3.2 again gives the reduced energy expression (2.129 and 2.130), i.e.,

$$X = (E - \alpha)/2\beta = \cos\theta. \tag{4.4}$$

The general solution of (4.1) can be written as (cf. 2.132)

$$c_n = a\ e^{in\theta} + b\ e^{-in\theta} \tag{4.5}$$

which by Euler's formula becomes

$$c_n = A\cos n\theta + B\sin n\theta. \tag{4.6}$$

With the aid of (4.4) and (4.5), equations (4.2) and (4.3) lead to

$$\begin{gathered} a(ze^{i\theta} - 1) + b(ze^{-i\theta} - 1) = 0 \\ a\ e^{iN\theta}(z - e^{i\theta}) + b\ e^{-iN\theta}(z - e^{-i\theta}) = 0 \end{gathered} \tag{4.7}$$

where

$$z = (\alpha' - \alpha)/\beta \tag{4.8}$$

is the *dimensionless surface perturbation parameter.* Eliminating a and b in (4.7) gives (Mark 1970)

$$\sin(N+1)\theta - 2z\sin N\theta + z^2\sin(N-1)\theta = 0. \tag{4.9}$$

It is worth noting that, on setting $z = \xi/R$, this equation becomes *identical* to (3.28) for the KP crystal. Since the solutions of such an equation have been discussed already in §3.3 for a finite crystal, here we shall consider only the case of a quasi-finite crystal, where N is very large.

4.1.1 Quasi-finite chain of atoms

By taking N very large, the two end effects can be isolated, so that only the perturbation at $n = 1$ need be considered. In this situation, (4.3) is replaced by (Grimley 1960)

$$c_{N+1} = 0 \tag{4.10}$$

which, as in (3.40), amounts to neglecting the surface effect at $n = N(\to \infty)$. Imposing (4.10) on (4.6) yields[1]

$$c_n = \frac{A}{\sin(N+1)\theta} \sin(N+1-n)\theta; \quad n = 1, 2, \ldots, N. \tag{4.11}$$

Inserting (4.11) in (4.2), and using (4.4) and (4.8), leads to the equation (cf. 3.41)

$$z = \frac{\sin(N+1)\theta}{\sin N\theta} \tag{4.12}$$

whose roots give the values of θ. The graphical solution of (4.12) is displayed in Fig. 4.2. The curves of the right-hand side of (4.12) have asymptotes at $\theta = j\pi/N$, where $j = 1, 2, \ldots, N$. The intercepts on the ordinate axes at $\theta = 0$ and π are $(1 + N^{-1})$ and $-\ (1 + N^{-1})$, which, as $N \to \infty$, become 1 and -1, respectively. Thus, for any horizontal line in the range $-1 < z < 1$, there are N real roots, as indicated, each corresponding to a real value of θ. Such real solutions have reduced energies X lying in the bulk band $-1 < X < 1$ via (4.4), and delocalized periodic wave functions corresponding to (4.11) spread throughout the chain. When $z > 1$ ($z < -1$), a real root disappears at $\theta = 0$ ($\theta = \pi$), and only $(N - 1)$ real roots remain.

Analytically, the missing solution can be investigated by taking θ *complex* and writing

$$\theta = \zeta + i\mu \tag{4.13}$$

μ being real and positive. In this case, (4.4) becomes

$$X = \cos\zeta \cosh\mu - i \sin\zeta \sinh\mu. \tag{4.14}$$

However, since X must always be real, the imaginary part in (4.14) must be zero. Since $\mu > 0, \sinh\mu \neq 0$, so $\sin\zeta = 0$, which means that

$$\zeta = m\pi; \quad m = 0, 1, \ldots \tag{4.15}$$

[1]Solutions (4.11) are of the form of Chebyshev polynomials of the second kind, i.e., $U_n(\cos\theta) = \sin(n+1)\theta/\sin\theta$, which is not surprising, since (4.1) is their recurrence relation (Abramowitz and Stegun 1972).

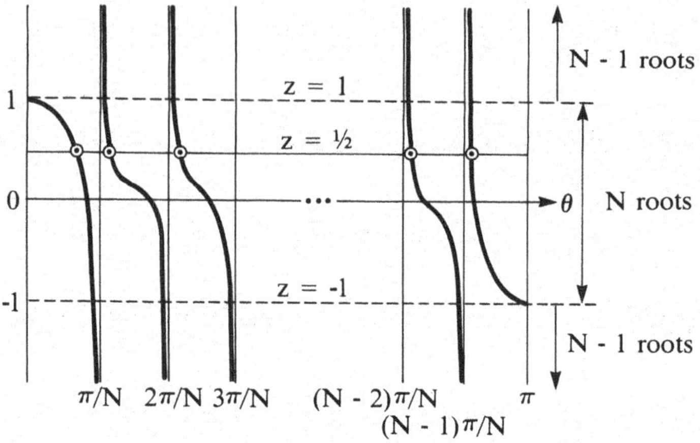

Fig. 4.2. Graphical solution of $z = \sin(N+1)\theta/\sin N\theta$.

Here, only the values of $m = 0$ and 1 are required, because all other values merely repeat these solutions. Hence, (4.13) gives the θ-values

$$\theta = i\mu, \quad \theta = \pi + i\mu \tag{4.16}$$

for the two missing solutions at $\theta = 0$ and π in Fig. 4.2.

(i) *P-state*

When $\theta = i\mu$, equation (4.4) shows that

$$X = \cosh \mu. \tag{4.17}$$

Since $\cosh \mu$ is always *positive*, this state is called a *P-state.* Moreover, since $X > 1$, the energy of the P-state lies *above* the upper band edge of the bulk states. Putting $\theta = i\mu$ in (4.11), and taking $N \to \infty$, leads to

$$c_n = A\, e^{-n\mu}. \tag{4.18}$$

Thus, the wave function of the P-state is exponentially damped from the surface into the chain. Such a state is called a *localized Tamm surface P-state.* Proceeding in the same way, (4.12) reduces to

$$z = e^{\mu} \tag{4.19}$$

which, for $\mu > 0$, gives (Fig. 4.2)

$$z > 1 \tag{4.20}$$

as the P-state *existence condition.* It is evident from (4.19) that μ is determined by the value of the surface perturbation parameter z. Using (4.19), equations (4.17) and (4.18) can be expressed in terms of z, namely,

$$X = \tfrac{1}{2}\,(z + z^{-1}) \tag{4.21}$$

and

$$c_n = A\ z^{-n}. \tag{4.22}$$

(ii) *N-state*

For $\theta = \pi + i\mu$, equation (4.4) gives

$$X = -\cosh\mu \tag{4.23}$$

so X is always *negative* and an *N-state* with $X < -1$ lies *below* the lower bulk-band edge. Equations (4.18), (4.19) and (4.20) are now replaced by

$$c_n = A(-1)^n\ e^{-n\mu} \tag{4.24}$$

$$z = -\ e^{\mu} \tag{4.25}$$

and

$$z < -1 \tag{4.26}$$

respectively. Consequently, the N-state wave function is an exponentially-damped oscillatory function, which decays from the surface into the chain and characterizes a *localized Tamm surface N-state*, whose existence condition is (4.26). Note that (4.25) in (4.23) and (4.24) again yields (4.21) and (4.22) for the $X(z)$ and $c_n(z)$ expressions.

Summarizing, for $z = 0$, the chain has N delocalized states (4.11) lying in a reduced energy band (4.4) of width $\Delta X = 2$. As z increases above zero ($\alpha' > \alpha$ (4.8), so the surface atom is electron repelling), the energies of all the band states (4.4) increase slightly (the maximum amount being proportional to N^{-1}), until at $z = 1$ a P-type surface state separates out above the band leaving $(N-1)$ delocalized states in the band. X increases steadily with z, and the wave function (4.22) becomes more concentrated on the $n = 0$ atom (Grimley 1960). Conversely, as z decreases below zero

($\alpha' < \alpha$ (4.8), so the surface atom is electron attracting), the band-state energies decrease, until at $z = -1$ an N-type surface state separates out below the band, leaving $(N-1)$ delocalized states in the band. As z decreases further, the N-state falls further below the band, and the wave function becomes more localized on the surface atom. The above findings are illustrated schematically in Fig. 4.3.

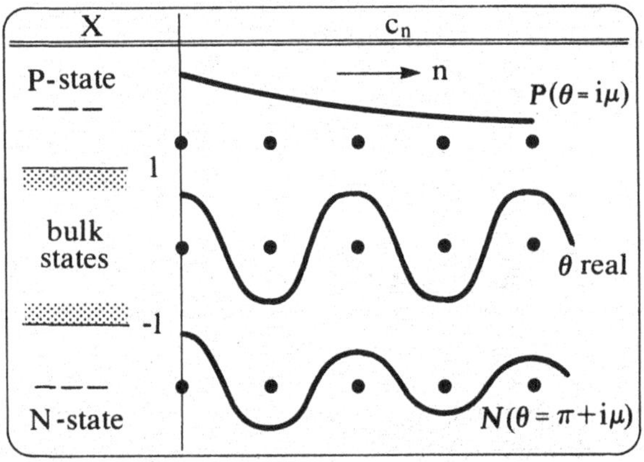

Fig. 4.3. Display of energy spectrum and wave-function variation.

4.2 Tamm States of a Cubic Crystal

The above procedure, for a one-dimensional crystal, can be generalized directly to the case of a three-dimensional crystal for a *simple cubic crystal*, where the separation of variables enables the Schrödinger equation to be written for each direction. Here, the three-dimensional Hamiltonian (wave function) can be expressed as a sum (product) of the Hamiltonians (wave functions) in each of the three directions, i.e.,

$$H = \sum_{i=1}^{3} (-\nabla_i^2 + V_i) = \sum_{i=1}^{3} H_i \tag{4.27}$$

and

$$\psi(x_1, x_2, x_3) = \prod_{i=1}^{3} \psi_i(x_i) \tag{4.28}$$

respectively. Thus, inserting (4.27) and (4.28) in the three-dimensional Schrödinger equation

$$H\psi = E\psi \tag{4.29}$$

shows that

$$H_i\psi_i(x_i) = E_i\psi_i(x_i) \tag{4.30}$$

where

$$E = \sum_{i=1}^{3} E_i. \tag{4.31}$$

Consider now a simple cubic lattice defined by the three integers (n_1, n_2, n_3), which take the values $0, 1, \ldots, N-1$. Let the plane $n_1 = 0$ define the surface (100) on which all the Coulomb integrals α are changed to α' (Goodwin 1939b, Baldock 1952, Grimley 1960). Assuming the wave functions vanish on the other five surfaces of the cube, then, in accordance with (4.11) and (4.28), the *unnormalized* wave-function coefficients are given by

$$c(n_1, n_2, n_3) = \sin(N - n_1)\theta_1 \sin n_2\theta_2 \sin n_3\theta_3 \tag{4.32}$$

where θ_1 is one of the N roots of (4.12), and $\theta_j = m_j\pi/(N-1)$; $m_j = 1, 2, \ldots, N-1$; $j = 2$ and 3. ($m_j = 0$ gives the physically uninteresting results $c_{m_1 00}(n_1, n_2, n_3) \equiv 0$ and this is due to a particular non-symmetrical choice of the origin of the rectangular coordinate system.) From (4.4) and (4.31), it follows that

$$X(\theta) = \sum_{i=1}^{3} \cos\theta_i \tag{4.33}$$

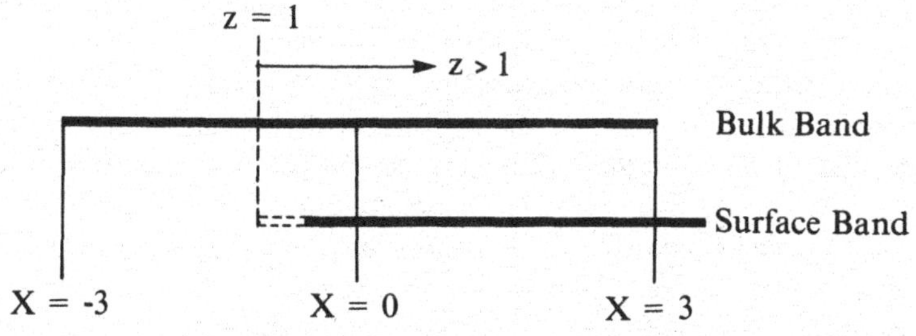

Fig. 4.4. Overlap and separation of bulk and surface bands.

so the *bulk bandwidth* is

$$\Delta X_b = |X(0) - X(\pi)| = 6. \tag{4.34}$$

For surface states, (4.21) and (4.33) yield

$$X(z,\theta) = \tfrac{1}{2}(z + z^{-1}) + \sum_{i=2}^{3} \cos\theta_i. \tag{4.35}$$

Thus, the *surface bandwidth* is

$$\Delta X_s = |X(z,0) - X(z,\pi)| = 4. \tag{4.36}$$

Comparing (4.33) with (4.35) reveals that, for $z \simeq 1$, the bulk band overlaps the surface one, but, as z increases, the two bands start to separate (Fig. 4.4). Note that, while the separation of the bulk and surface bands depends on z, their bandwidths do not. Hence, increasing z above unity causes the surface band to be rigidly shifted to the right, so that it begins to separate from the upper edge of the bulk band. This is analogous to the emergence of the P-state from the band in the one-dimensional treatment. In the three-dimensional situation, the surface band contains N^2 levels, as opposed to the single discrete state of the one-dimensional case. The N^2 levels are, of course, missing from the bulk band.

4.3 Shockley States of an sp-Hybrid Chain

Although Tamm's discovery of surface states in 1932 sparked a flurry of theoretical research activity in the ensuing years (Fowler 1933, Rijanov 1934, Sokolov 1934, Maue 1935 and Goodwin 1939), it was Shockley (1939) who studied surface states from a somewhat different perspective by investigating how their energy levels originated from the discrete levels of atoms that coalesced from an infinite separation to form a crystal with a finite lattice spacing. In order to trace the evolution of the energy spectrum as the lattice spacing was reduced, Shockley used a method, which was later developed into the *self-matching cell solutions* approach (James 1949), to solve the Schrödinger equation for a *finite* arbitrary crystal potential.

For pedagogical reasons, it is easier to discuss Shockley states within the framework of the TBA, which enables the required *band-crossing* feature of the original work to be modelled in a simple way. In §4.1, the electronic structure of the crystal consisted of one band derived from a single atomic p-orbital. For band-crossing purposes, a minimum of two orbitals per atom is needed. Consequently, an s- and p-orbital are associated with each atom in the chain (Fig. 4.5). These orbitals can be combined to form two *sp-hybrid* orbitals in the elementary cell of the infinite chain.

Each hybrid orbital can now be taken as a one-electron orbital around an atomic site with a Coulomb integral α. If the interaction between the hybrids on each atom is weaker than that between hybrids on adjacent atoms, then the sp-hybrid chain is equivalent to a monatomic one with alternating weak (β) and strong (β') bonds (Fig. 4.5). Such a situation is found in long-chain polyenes (Tsuji *et al.* 1960).

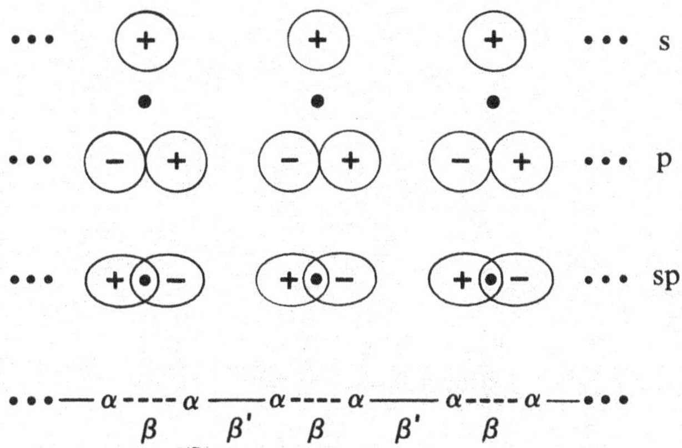

Fig. 4.5. Model of sp-hybrid chain.

Koutecký (1961) showed that Shockley states arise when a semi-infinite crystal is formed by breaking the *strong* bonds between atoms *without* perturbing the surface one. Contrary to his next-nearest-neighbour treatment, the nearest-neighbour one is adopted here, for simplicity. For the monatomic equivalent of the sp-hybrid chain, containing N (even) atoms numbered 0 to $N-1$, equation (4.1) takes the reduced forms

$$\begin{aligned} Xc_n &= c_{n+1} + \eta c_{n-1}, \qquad n \text{ even} \\ Xc_n &= \eta c_{n+1} + c_{n-1}, \qquad n \text{ odd} \end{aligned} \tag{4.37}$$

where[2]

$$X = (E-\alpha)/\beta, \quad \eta = \beta'/\beta. \tag{4.38}$$

Solutions of (4.37) are sought by taking

[2]Note that X replaces $2X$ of (4.4).

$$
\begin{aligned}
c_n &= u^n, \quad n \text{ even} \\
c_n &= v^n, \quad n \text{ odd.}
\end{aligned}
\tag{4.39}
$$

Inserting (4.39) in (4.37), and eliminating v, leads to the quadratic (in u^2) equation

$$\eta\, u^4 + (1 + \eta^2 - X^2)u^2 + \eta = 0 \tag{4.40}$$

whose roots are

$$u^2 = (2\eta)^{-1} \left\{ (X^2 - 1 - \eta^2) \pm [(1 + \eta^2 - X^2)^2 - 4\eta^2]^{1/2} \right\}. \tag{4.41}$$

By choosing

$$X = \pm(1 + \eta^2 + 2\eta \cos 2\theta)^{1/2} \tag{4.42}$$

(4.41) reduces to

$$u^2 = e^{\pm 2i\theta}$$

so the general coefficient (4.39) for n *even* is (cf. 4.6)

$$c_n = A \cos n\theta + B \sin n\theta. \tag{4.43}$$

Applying the boundary condition $c_N = 0$ (4.10) for N even to (4.43) gives (cf. 4.11)

$$c_n = \frac{A}{\sin N\theta} \sin(N - n)\theta; \quad n = 0, 2, \ldots, N - 2. \tag{4.44}$$

For n *odd*, (4.44) in (4.37) yields the coefficient expression

$$Xc_n = \frac{A}{\sin N\theta} [\eta \sin(N - n - 1)\theta + \sin(N - n + 1)\theta]; \quad n = 1, 3, \ldots, N - 1 \tag{4.45}$$

which satisfies the boundary condition

$$Xc_{N-1} = c_{N-2}. \tag{4.46}$$

Equation (4.42) shows that the bulk-band edges arise for $\theta = 0$ and $\pi/2$, which lead to the straight-line band-edge expressions

$$X = \pm(1 \pm \eta). \tag{4.47}$$

The linear variations of X with η are displayed in Fig. 4.6. For $\eta = 0$ ($\beta' = 0$), there is no inter-atomic interaction between the s- and p-orbitals, so $X = +1(-1)$ corresponds to the energy of an isolated atom's p(s)-state.[3] As η increases, the interaction between the s- and p-states increases, causing the discrete states to broaden out into s- and p-type bulk bands (shaded areas). At $\eta = 1$, the bands cross, giving rise to the so-called *direct* gap ($\eta < 1$) and the *inverted* gap ($\eta > 1$). The former is bounded above (below) by the p-band (s-band), while the reverse is true in the latter case. It is also noteworthy that, at $\eta = 1$ (i.e., $\beta' = \beta$), the sp chain reduces to a monatomic one, which is reflected in Fig. 4.6 by the coalescing of the two bulk bands into a single band.

Returning to (4.37), the surface equations for the two atoms in the first unit cell are

$$Xc_0 = c_1 , \qquad n = 0 \tag{4.48}$$

$$Xc_1 = \eta c_2 + c_0 , \quad n = 1. \tag{4.49}$$

Eliminating c_1 in these equations gives

$$X^2 - 1 = \eta c_2 / c_0 \tag{4.50}$$

[3]If, in the Dirac (1958) $\langle$bra|c|ket$\rangle$ notation, the kets $|s\rangle$ and $|p\rangle$ denote the s- and p-atomic orbitals, then their corresponding energies are

$$\langle s|H_a|s\rangle = \epsilon_s , \qquad \langle p|H_a|p\rangle = \epsilon_p$$

where H_a is the atomic Hamiltonian. The sp-hybrid orbitals

$$|\psi_\pm\rangle = (|p\rangle \pm |s\rangle)/\sqrt{2}$$

now lead to

$$\langle\psi_\pm|H_a|\psi_\pm\rangle = (\epsilon_s + \epsilon_p)/2 = \alpha$$

$$\langle\psi_+|H_a|\psi_-\rangle = (\epsilon_p - \epsilon_s)/2 = \beta$$

since $\langle s|H_a|p\rangle = \beta' = 0$. Thus,

$$\alpha - \beta = \epsilon_s , \qquad \alpha + \beta = \epsilon_p$$

which, in reduced form, give

$$(\epsilon_{p,s} - \alpha)/\beta = X_{p,s} = \pm 1.$$

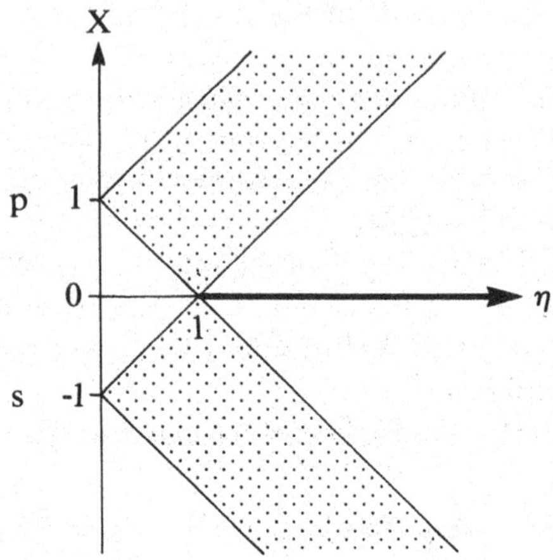

Fig. 4.6. Crossing of s- and p-bands in X vs. η plane, giving rise to direct ($\eta < 1$) and inverted ($\eta > 1$) gaps. Solid line $X = 0$ for $\eta > 1$ denotes Shockley state. Note, $\eta < 0$ case, is reflection of figure in the X-axis.

which by (4.42) and (4.44) leads to the θ-value equation

$$-\eta = \frac{\sin(N+2)\theta}{\sin N\theta} \tag{4.51}$$

for $0 \leq \theta \leq \pi/2$. However, since N is even (i.e., $N = 2M$, say), this can be written as

$$-\eta = \frac{\sin(M+1)\phi}{\sin M\phi} \tag{4.52}$$

where $0 \leq \phi = 2\theta \leq \pi$. On replacing $-\eta$ with z, equation (4.52) becomes identical to (4.12), whose graphical solutions are shown in Fig. 4.2. Thus, for $|\eta| \leq 1$, (4.51) has $N/2$ *real* solutions of $\theta = [0, \pi/2]$ in each band of Fig. 4.6. When $\eta < -1$ ($\eta > 1$), a real solution is lost at $\theta = 0$ ($\theta = \pi/2$) and becomes *imaginary* (*complex*), i.e., $\theta = i\mu$ ($\theta = \pi/2 + i\mu$). For these θ-values, the $N \to \infty$ limit of (4.51) is

$$-\eta = \pm\, e^{2\mu} \gtrless \pm 1. \tag{4.53}$$

Correspondingly, (4.42) gives the energies of these so-called *Shockley* surface states, viz.,

$$X = \pm(1 + \eta^2 \pm 2\eta \cosh 2\mu)^{1/2} \tag{4.54}$$

which by (4.53) reduce to the doubly-degenerate level

$$X = 0 \tag{4.55}$$

in the inverted gap of Fig. 4.6. In the case of n even, the surface-state wave function, obtained from (4.44) for $N \to \infty$, is (cf. 4.18 and 4.24)

$$\begin{aligned} c_n &= A\, e^{-n\mu}\,, & \theta &= i\mu \\ c_n &= A(-1)^{n/2}\, e^{-n\mu}, & \theta &= \pi/2 + i\mu. \end{aligned} \tag{4.56}$$

For n odd, (4.55) and (4.48) show that $c_1 = 0$, while (4.37a) then reveals that all subsequent odd coefficients vanish, in accord with Koutecký (1961).

Note that Shockley states are due solely to the effect of terminating the sp-hybrid chain, and do not involve the perturbing of the end atom, as in the Tamm-state situation. Needless to say, it is a simple matter to construct a model which enables both Tamm- and Shockley-state features to be analysed together by combining the two approaches described here. Such a model has been investigated by Foo and Wong (1974), who used a Green-function technique (§8.2).

5

NEARLY-FREE ELECTRON APPROACH

So free we seem, so fettered fast we are.
— Robert Browning

Although Maue (1935) was the first to use the *nearly-free electron approximation* (NFE) to study surface states, and their effect on the electrical conductivity of a metal, it was the subsequent generalization of the method by Goodwin (1939a) that eventually brought it into prominence. Unlike Goodwin's three-dimensional treatment, a one-dimensional one is developed here, for the sake of simplicity.

5.1 Complex Wave Number

As discussed in Chapters 3 and 4, k may assume *complex* values, which give rise to decaying wave functions, and energy levels that lie in the gaps between the k-real bands of the infinite crystal. Thus, the energy E vs k diagram has a k *real* band alternating with a k *complex* gap (FEG). It is now possible to trace the variation of E with k complex, i.e.,

$$k = \zeta + i\mu, \qquad \mu \geq 0 \tag{5.1}$$

continuously across the entire energy spectrum of the reduced zone scheme (Fig. 5.1). Starting at the origin and moving to A requires $\mu = 0$ in (5.1), while ζ varies from zero to π/a. Along the path ABC, $\zeta = \pi/a$, and μ goes from zero to μ_m and back to zero again. The energy is increasing and real over all this route, which passes through the lower band and first FEG to the lower edge of the second band. Since E remains a real function of a complex k on ABC, such lines are called *real lines* (Heine 1963, 1964). All pairs of bands are linked together by these loops of real lines for which k is complex. In this way, the energy can be traced monotonically from

$E = -\infty$ to $E = +\infty$, each turning point in the real band structure being a saddle point of E as a function of complex k (Heine 1964, Jones 1975).

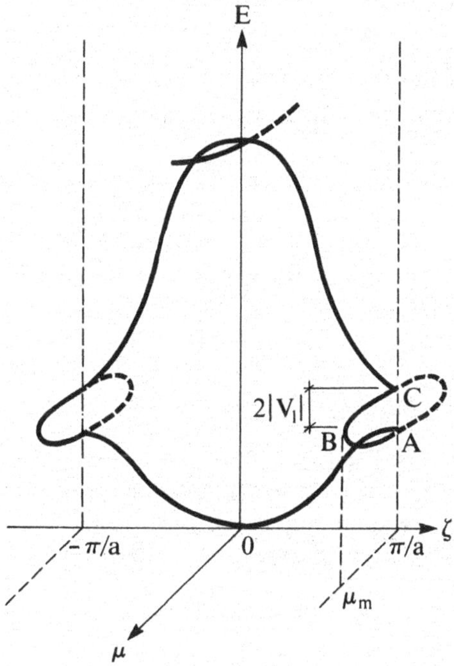

Fig. 5.1. Variation of energy E in complex plane (ζ, μ).

The above description of the variation of E with k complex can be placed on a more mathematical footing by considering the roots of (2.182), viz.,

$$E_k^{\pm} = \tfrac{1}{2}\left\{(k^2 + k_b^2) \pm [(k^2 - k_b^2)^2 + 4|V_g|^2]^{1/2}\right\} \tag{5.2}$$

where

$$k_b = k + 2\pi b_g. \tag{5.3}$$

For E to be real, the discriminant in (5.2) must be zero or positive. This places limits k_m on k, which are given by

$$(k_m + \pi b_g)^2 = -(V_g/2\pi b_g)^2. \tag{5.4}$$

Inserting (5.1) in (5.4), and equating the real parts and then the imaginary parts of each side of the resulting equation, shows that

$$\begin{aligned} \zeta_m &= -\pi b_g = \mp\pi g/a \\ \mu_m &= \pm|V_g|a/2\pi g \end{aligned} \tag{5.5}$$

by (2.180). Equations (5.5) locate the limits of k on the real lines in each of the gaps (Fig. 5.1), the corresponding energy being

$$\begin{aligned} E_k^m &= \tfrac{1}{2}(k_m^2 + k_b^2) = \zeta_m^2 - \mu_m^2 \\ &= (\pi g/a)^2 - (V_g a/2\pi g)^2. \end{aligned} \tag{5.6}$$

To obtain the wave function, it is necessary to return to (2.188) and realize that

$$\begin{aligned} \cos(\pi x/a) &= Re\left(e^{\pm i\pi x/a}\right) \\ \sin(\pi x/a) &= Re\left[e^{\pm i(\pi x/a - \pi/2)}\right]. \end{aligned} \tag{5.7}$$

Thus, in crossing the *entire* gap from one band edge to the other, a phase factor of $-\pi/2$ is introduced into the wave function. Therefore, for any energy *in* the FEG, a phase factor δ ($|\delta| \leq \pi/2$) appears in the wave function expression. Consequently, for the first FEG, $g = 1$ in (5.5), thus

$$k = i\mu \pm \pi/a$$

via (5.1) with μ to be determined. Assuming

$$k = i\mu + \pi/a, \qquad \mu \geq 0 \tag{5.8}$$

i.e., the reciprocal vector π/a in the direction of the normal to the crystal surface at $x = 0$, leads to the following form of the wave function

$$Re\left\{e^{-i[(i\mu+\pi/a)x+\delta]}\right\}$$

i.e.,

$$\psi(x) = e^{\mu x}\cos(\pi x/a + \delta), \qquad x \leq 0. \tag{5.9}$$

Hence, like the energy, the wave function is now analytically continuous across the FEG. Actually, (5.9) represents a 'mixture' of the wave functions $\psi^{\pm}$ (2.187) associated with the upper and lower FEG edges, the continuously varying phase shift δ being the *mixing* parameter. It should be remembered that (5.9) is the wave function within the so-called *two-band* model, i.e., it has been derived by neglecting the influence of other energy bands (cf. 2.170).

On writing

$$\cos(\pi x/a + \delta) = \tfrac{1}{2}e^{i\delta}\left(e^{i\pi x/a} + e^{-2i\delta}e^{-i\pi x/a}\right) \tag{5.10}$$

then, since the factor $\frac{1}{2}\, e^{i\delta}$ can be absorbed in the normalization constant, comparing (5.10) with (2.178) reveals that, for the first FEG,

$$e^{2i\delta} = (E - k^2)/V_1\ . \tag{5.11a}$$

From (5.11a), it is evident that the sign of V_1 is important. Let us therefore explore briefly the mathematical significance and physical consequences of the sign of V_1, the Fourier expansion component in (2.145). Since $V(\mathbf{r})$ is real in (2.145), $V_{\mathbf{g}} = V^*_{-\mathbf{g}}$, which, for a one-dimensional two-band model, leads to

$$V(x) = 2V_1 \cos(2\pi x/a)$$

since $V_0 = 0$ via (2.147). However, it is customary (cf. Forstmann 1978) to adopt Goodwin's (1939a) sign convention and write

$$V(x) = -2V_1 \cos(2\pi x/a). \tag{5.11b}$$

We shall henceforth follow this convention. Now, the condition $V_1 > 0$ means that the expansion at $x = 0$ is about a relative minimum, as in Fig. 5.2, where the origin is located at the centre of an atom.

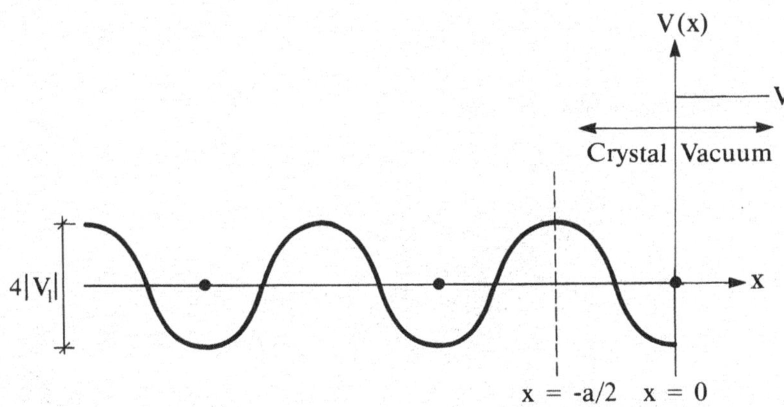

Fig. 5.2. NFE crystal terminated by constant potential V_0 in vacuum region $(x > 0)$.

Let us return to (5.11a), which, under the assumed sign convention (5.11b), takes the form

$$e^{2i\delta} = -(E - k^2)/V_1 \ . \tag{5.11c}$$

Remembering E is real, inserting (5.8) in (5.11c) and equating the imaginary parts gives

$$\sin 2\delta = 2\pi\mu/V_1 a \tag{5.12}$$

while the real parts yield

$$\cos 2\delta = -(E - \pi^2/a^2 + \mu^2)/V_1. \tag{5.13}$$

Equation (5.12) gives the relationship between the ranges of δ and the sign of V_1, while (5.13) determines the values of δ for the lower and upper FEG edges. Squaring and adding (5.12) and (5.13) leads to

$$\mu^2 = -E - \pi^2/a^2 + (V_1^2 + 4\pi^2 E/a^2)^{1/2}. \tag{5.14}$$

Note that μ reaches its limits (5.5) at E given by (5.6) and

$$\mu = 0 \tag{5.15}$$

at the FEG edges (2.188). Since μ is real, the right-hand side of (5.14) must be positive, which requires the energy to lie in the gap defined by $(\pi/a)^2 - |V_1| < E < (\pi/a)^2 + |V_1|$ (cf. 2.188), as expected. Assuming

$$\mu \geq 0 \tag{5.16}$$

(5.12) gives the following relations

$$V_1 < 0, \qquad -\pi/2 \leq \delta \leq 0 \tag{5.17}$$

$$V_1 > 0, \qquad 0 \leq \delta \leq \pi/2. \tag{5.18}$$

It should be noted that the upper (E^+) and lower (E^-) FEG edges are

$$E^{\pm} = \pi^2/a^2 \pm |V_1| \tag{5.19}$$

by (2.186), and $\mu(E^{\pm}) = 0$ from (5.15), so for $V_1 < 0$ (5.13) gives

$$\cos 2\delta(E^{\pm}) = \mp|V_1|/V_1 = \pm 1 \ . \tag{5.20}$$

Thus, $\delta(E^+) = 0$ and $\delta(E^-) = -\pi/2$ in (5.9) give the wave functions for the upper (ψ^+) and lower (ψ^-) FEG edges, which are even (cosine) and odd (sine) respectively, i.e., exhibit s- and p-like character.

On the other hand, for $V_1 > 0$, $0 \leq \delta \leq \pi/2$, so (5.13) gives

$$\cos 2\delta(E^{\pm}) = \mp 1. \tag{5.21}$$

Therefore, for $V_1 > 0$, the symmetry of the wave functions, bounding the FEG, are the reverse of those originating from $V_1 < 0$ (2.188)[1]. That is, the wave function for the lower FEG edge is even (s-type), while that for the upper FEG is odd (p-type), as follows from putting $\delta = 0$ and $\pi/2$ in (5.9) with $\mu = 0$. Such a FEG is known as the *direct* gap (Forstmann 1970, 1978). Conversely, the FEG with the p-band edge at the bottom is often called an *inverted* gap to stress the inverted s-p order of the isolated atomic levels (cf. §4.3).

5.2 Surface-State Analysis

As shown in §5.1, if a surface state exists in the FEG, then its wave function inside the semi-infinite crystal must have the form

$$\psi_c(x) = \beta e^{\mu x} \cos(\pi x/a + \delta), \qquad x \leq 0 \tag{5.22}$$

with δ and μ given by (5.12) and (5.14), respectively. For $\mu > 0$ and $x < 0$, (5.22) represents the exponentially decaying oscillatory wave function *inside* the crystal, i.e., the so-called *evanescent* solution. The next step is to check if (5.22) can be matched to the wave function *outside* the crystal. If a vacuum of constant potential V_0 is present outside the crystal, then

$$\psi_v(x) = \alpha e^{-k_0 x}, \qquad x \geq 0 \tag{5.23}$$

$$k_0 = (V_0 - E)^{1/2}, \qquad E < V_0. \tag{5.24}$$

In general, the matching can be performed at arbitrary $x = x_0$ ($x_0 < 0$ – the position of the crystal surface). Now, if the logarithmic derivatives of $\psi_c(x)$ and $\psi_v(x)$ can be matched at $x = x_0$, then the wave function of the system will decay in *both* directions from the surface, and the electron will be *bound* (or localized) at the surface, i.e., will be in a *surface state.* Performing this matching procedure leads to an equation whose solution

[1]Note, in order to use (2.188), the transformation $V_1 \to -V_1$ must be made, in accordance with (5.11b).

gives the surface-state energy in the FEG. Conversely, the surface-state energy equation can be regarded as a restrictive condition, which determines the energy at which a continuous matching of the wave functions and their derivatives is possible.

Matching $\psi'(x)/\psi(x)$ of (5.22) and (5.23) at $x = x_0$ leads to

$$(\pi/a)\tan(\pi x_0/a + \delta) = (\mu + k_0) \tag{5.25}$$

which, is the equation for the *surface-state energy* E_s, since δ (5.12), μ (5.14) and k_0 (5.24) are energy dependent.

Two special cases of (5.25) are of particular interest, i.e.,

$$x_0 = 0 \Longrightarrow (\pi/a)\tan\delta = (\mu + k_0) \tag{5.26}$$

$$x_0 = -a/2 \Longrightarrow (\pi/a)\cot\delta = -(\mu + k_0). \tag{5.27}$$

Since, according to (5.16) and (5.24), μ and k_0 are always positive, (5.26) and (5.27) can be satisfied only for $0 \leq \delta \leq \pi/2$ or $-\pi/2 \leq \delta \leq 0$, respectively. Consequently, the necessary condition for a surface state to appear in the FEG is $V_1 < 0$ for $x_0 = -a/2$ and $V_1 > 0$ for $x_0 = 0$ (cf. 5.17 and 5.18). This very interesting and important result was noted first by Maue (1935) and discussed subsequently by Goodwin (1939a)[2], Forstmann (1970, 1978), Jones (1975) and Pendry and Gurman (1975).

Consider first the matching at $x = 0$ (here, and henceforth, the zero subscript is omitted, for convenience). The solution of (5.26) gives the energy E_s of a surface state, which appears in a direct FEG with the s-band at the bottom, as immediately follows from (5.22), (5.21) and (5.15). Now, the condition that such a state can exist only if $V_1 > 0$, can be visualized in the following way. The wave functions for the FEG edges $\psi^+(x) \sim \sin\pi x/a$ and $\psi^-(x) \sim \cos\pi x/a$ have zero value and zero slope at $x = 0$, respectively, so they can never be matched to the vacuum wave function (5.23) for which $\psi_v(0) > 0$ and $\psi_v'(0) < 0$. For $0 < \delta < \pi/2$, it follows from (5.22) that $\psi_c(0) > 0$. Now, when (5.26) is satisfied, the inequality

$$(\pi/a)\tan\delta > \mu \tag{5.28}$$

is obviously valid. However, (5.28) is equivalent to $\psi_c'(0) < 0$, so the maxima of $\psi_c(x)$ must be shifted to the left, away from the crystal surface at

[2]Maue (1935) and Goodwin (1939) considered only the case when the crystal was terminated at $x = 0$.

$x = 0$, for a smooth matching to occur (Fig. 5.3 – solid line). Consequently, the presence of a δ *phase-shift* in the surface-state wave function (5.22) *raises* the energy from the band edge E^- to E_s in the FEG, as is clearly seen in Fig. 5.4, where a schematic plot of (5.26) is presented. A surface state occurring in a *direct* FEG with the s-band at the bottom is usually called a *Tamm* state. It should be noted that, for a negative δ phase-shift ($-\pi/2 < \delta < 0$), $\psi_c'(0) > 0$, so the maxima of $\psi_c(x)$ are shifted to the right, towards the surface, which means that a smooth matching of the wave functions and their derivatives at $x = 0$ is *not* possible (Fig. 5.3 – dashed line).

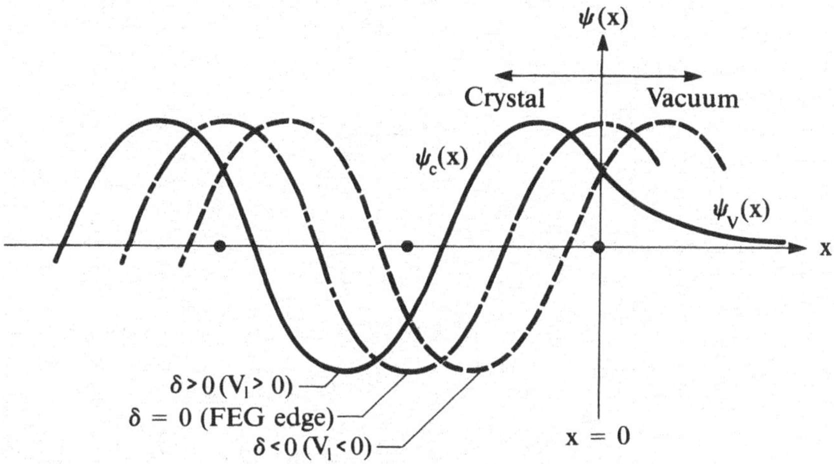

Fig. 5.3. (Non) Matching of wave function at crystal surface for phase shift $\delta(< 0) > 0$. Black dots depict positions of atoms.

Consider now the matching at $x = -a/2$, since, obviously, the origin ($x = 0$) and the surface do not have to coincide. In fact, in most calculations, the origin is located at an atom (i.e., $V_1 > 0$), while matching at the surface is taken at the midway position, i.e., at $x = -a/2$ (cf. Fig. 5.2). Clearly, matching at $x = -a/2$ is equivalent to the substitution

$$x = x' - a/2 \tag{5.29}$$

and matching at $x' = 0$. The corresponding wave functions for the upper

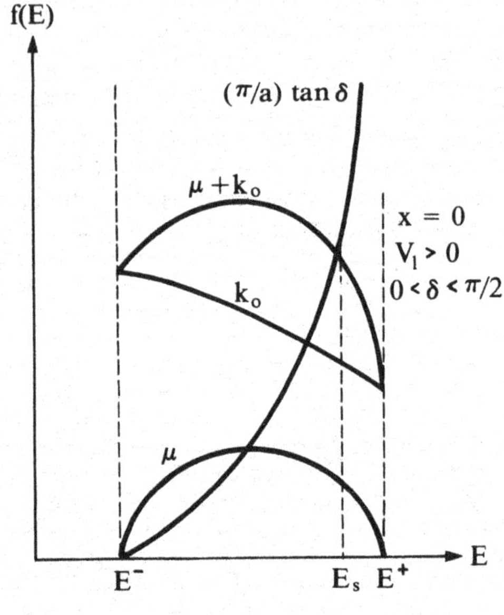

Fig. 5.4. Solution of surface-state energy equation (5.26).

($\delta(E^+) = \pi/2$) and lower ($\delta(E^-) = 0$) FEG edges (cf. 5.20) become

$$\psi^+(x') \sim \cos(\pi x'/a - \pi/2 + \pi/2) \sim \cos(\pi x'/a) \tag{5.30}$$

$$\psi^-(x') \sim \cos(\pi x'/a - \pi/2) \sim \sin(\pi x'/a) \; . \tag{5.31}$$

The resulting surface state with energy E_s, being a solution of (5.26), will occur, therefore, in an *inverted* gap with the p-band (5.31) at the bottom. Such a state is usually called a *Shockley state* (cf. §4.3), since Shockley (1939) was the first to introduce the idea of an inverted gap, when studying the potential with the termination midway between two atoms. Using a more general procedure than the NFE approximation, he formulated the criterion for the existence of surface states in terms of the inter-atomic distance (Fig. 5.5). When atoms are far apart, so that their wave functions do not overlap, their energy levels are discrete and are labelled in ascending order s, p, d, etc., levels. As the atoms coalesce, their wave functions overlap and bonding occurs between the atoms to form the crystal. The interaction between the atoms leads to a broadening of the discrete atomic levels into bands (Fig. 5.5). A further reduction of the lattice spacing results in a crossing of the s- and p-bands, after which a FEG is formed with upper and lower edges of inverted symmetry, so that the p-band is now the lower one[3]. According to Shockley's criterion, this is a necessary condition for surface states to exist, if the crystal is terminated midway between two atoms. Obviously, Shockley's criterion does not depend on the choice of the origin. Indeed, if the origin ($x = 0$) is in the midway position, then $V_1 < 0$ in (5.11b) (cf. Fig. 5.2) and the phase-shift δ varies from $-\pi/2$ (bottom of the FEG) to zero (top of the FEG) leading to an inverted FEG, as follows from (5.17) and (5.20).

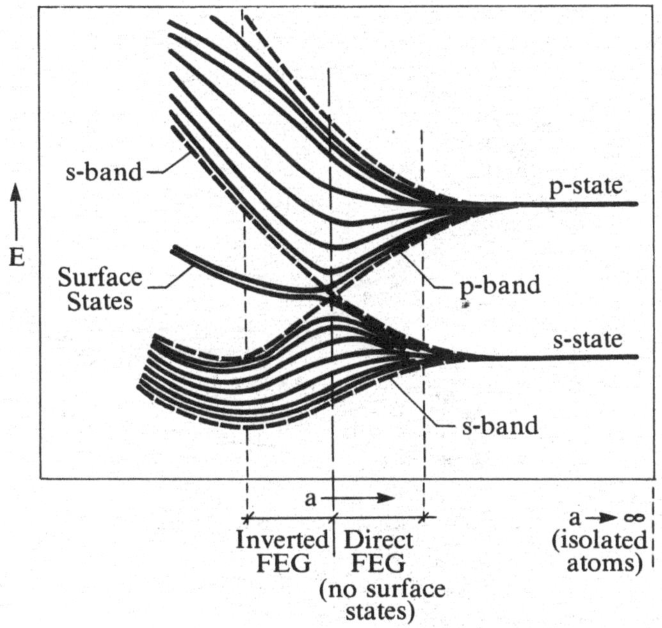

Fig. 5.5. Creation of s- and p-bands from isolated s- and p-atomic levels.

We can conclude, therefore, that the existence of a surface state in an inverted FEG, where the gap is p-like at the bottom and s-like at the top, requires $V_1 > 0$ ($0 < \delta < \pi/2$), if the origin $x = 0$ is at the centre of an

[3]Comparison of Fig. 5.5 with Fig. 4.6 shows that decreasing the lattice spacing is analogous to increasing η.

atom, and $V_1 < 0$ $(0 > \delta > -\pi/2)$ if it is put midway between the adjacent atoms.

5.3 Direct Application of Two-Band NFE Model

Despite its simplicity, the two-band NFE model provides a reasonably good description of a real band structure, as witnessed by its use in the study of surface states (in particular, image states) of fcc metals (Smith 1985, Bartynski *et al.* 1985). In Fig. 5.6, the band structures, derived from (5.2) and that obtained from realistic calculations for Cu(100) (Segall 1962), are presented. As can be seen, the appropriate choice of parameters V_g and a enables the model (Fig. 5.6a) to describe quite well the X gap of Cu(100) (Fig. 5.6b). Let us now briefly outline these model calculations.

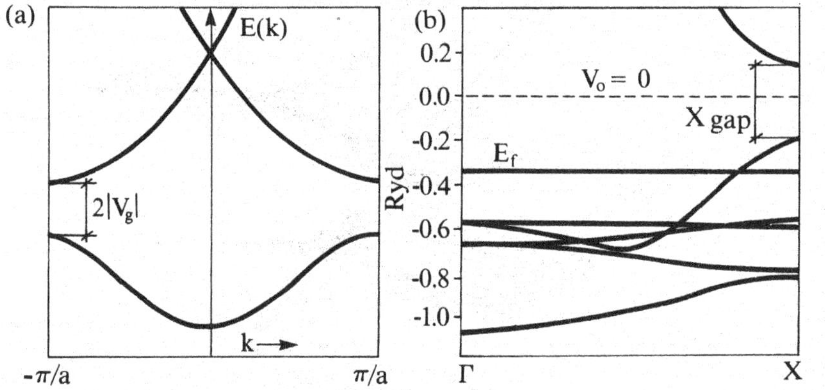

Fig. 5.6. (a) NFE band structure for one-dimensional crystal. (b) Real band structure for Cu(100) (Segall 1962).

5.3.1 (100) surface of fcc crystal

Using (2.146), Bloch functions (2.11) can be written as the plane-wave expansion

$$\psi_{\mathbf{k}}(\mathbf{r}) = \sum_{\mathbf{h}} a_{\mathbf{h}}\, e^{i(\mathbf{k}+2\pi\mathbf{b_h})\cdot\mathbf{r}} \,. \tag{5.32}$$

Since the components of the reciprocal vectors $\mathbf{g} = (g_1, g_2, g_3)$ and $\mathbf{h} = (h_1, h_2, h_3)$ are integers, $\mathbf{h}$ in (5.32) may be replaced by $\mathbf{h} - \mathbf{g}$ to give

$$\psi_{\mathbf{k}}(\mathbf{r}) = \sum_{\mathbf{h}-\mathbf{g}} a_{\mathbf{h}-\mathbf{g}}\, e^{i\left(\mathbf{k}-2\pi \mathbf{b}_{\mathbf{g}-\mathbf{h}}\right)\cdot\mathbf{r}} \,. \tag{5.33}$$

By separating out the different parallel components, (5.33), involving a finite number of terms, can be transformed into

$$\psi_{\mathbf{k}}(\mathbf{r}) = \sum_{m} a_m\, e^{i(k_x - 2\pi b_x^m)x}\, e^{i\mathbf{k}_{||}\cdot\mathbf{r}_{||}}\, e^{-i2\pi \mathbf{b}_{||}^m\cdot\mathbf{r}_{||}} \tag{5.34}$$

where the following notation has been used

$$\mathbf{k} = (k_x, \mathbf{k}_{||}),\ \mathbf{b}_{\mathbf{g}-\mathbf{h}} = (b_x^m, \mathbf{b}_{||}^m),\ \mathbf{r} = (x, \mathbf{r}_{||}), \quad m = 0, 1, 2, \ldots$$

Within the two-band NFE model, $m = 0$ and 1; assuming that [100] is along the x-axis gives, for a fcc crystal, $b_x^0 = 0, \mathbf{b}_{||}^0 = \mathbf{b}_{||}^1 = \mathbf{0}$, $b_x^1 = 2/a$. Thus, the crystal wave function (5.34) becomes

$$\psi_c(\mathbf{r}) = (a_0 e^{ikx} + a_1 e^{i(k-4\pi/a)x})e^{i\mathbf{k}_{||}\cdot\mathbf{r}_{||}} \tag{5.35}$$

where $k = k_x$.

Obviously, (5.35) must satisfy the Schrödinger equation

$$[-\nabla^2 + V(\mathbf{r})]\psi_c(\mathbf{r}) - E\psi_c(\mathbf{r}) = 0 \,. \tag{5.36}$$

Multiplying the resulting equation on the left, first by $e^{-ikx}\ e^{-i\mathbf{k}_{||}\cdot\mathbf{r}_{||}}$ and integrating over the unit cell, then by $e^{-i(k-4\pi/a)x}\ e^{-i\mathbf{k}_{||}\cdot\mathbf{r}_{||}}$ and again integrating over the unit cell, leads to the matrix equation for the coefficients a_0 and a_1, namely,

$$\begin{bmatrix} (k^2 + \mathbf{k}_{||}^2) - E & V \\ V^* & [(k - (4\pi/a))^2 + \mathbf{k}_{||}^2] - E \end{bmatrix} \begin{bmatrix} a_0 \\ a_1 \end{bmatrix} = 0 \tag{5.37}$$

where

$$V = \int V(\mathbf{r}) e^{-i(4\pi/a)x} d\mathbf{r} \Big/ \int d\mathbf{r} = -V_1 \tag{5.38}$$

with $V(\mathbf{r})$ given by (2.145). The requirement that the determinant of (5.37) be zero gives the relation

$$\{E - [(k - (4\pi/a))^2 + \mathbf{k}_{||}^2]\}[E - (k^2 + \mathbf{k}_{||}^2)] - |V|^2 = 0. \tag{5.39}$$

As expected, for $\mathbf{k}_{\|} = \mathbf{0}$, (5.39) reduces to (2.182) when $b_g = -(2/a)$. From (5.37) and (2.174), it follows that

$$\frac{a_1}{a_0} = \frac{E^{\pm} - (k^2 + \mathbf{k}_{\|}^2)}{V} = \pm\frac{|V|}{V} \tag{5.40}$$

where $V = -V_1$ and $+(-)$ sign corresponds to upper (lower) FEG limit.

For the first Brillouin zone edge, where the FEG is formed (Fig. 5.6a)

$$k = 2\pi/a \tag{5.41}$$

for the [100] direction in a fcc crystal. Inserting (5.40) and (5.41) into (5.35) gives, for $\mathbf{k}_{\|} = \mathbf{0}$,

$$\psi_c^{\pm}(x) = e^{i2\pi x/a} \mp \frac{|V_1|}{V_1}\, e^{-i2\pi x/a} \tag{5.42}$$

which coincides with (2.187). Therefore, further considerations are parallel to those presented in §5.2 and lead to the wave function of the form

$$\psi_c(x) \sim e^{\mu x} \cos[(2\pi x/a) + \delta], \quad x \leq 0 \tag{5.43}$$

with δ and μ given by

$$\sin 2\delta = 4\pi\mu/V_1 a \tag{5.44}$$

$$\mu^2 = -E - (2\pi/a)^2 + [V_1^2 + 4E(2\pi/a)^2]^{1/2} \tag{5.45}$$

(cf. 5.12 and 5.14).

6

MATHIEU POTENTIAL MODEL

What we experience of nature is in models, and all of nature's models are so beautiful.
— Buckminster Fuller

The first analysis of surface states on the basis of the Mathieu potential was provided by Statz (1950). However, because of the importance of the location of the surface in the first unit cell of the crystal, Levine (1968) carried out a more detailed analysis of the Mathieu problem for conditions of arbitrary termination. It is this latter approach which is adopted here.

6.1 Model

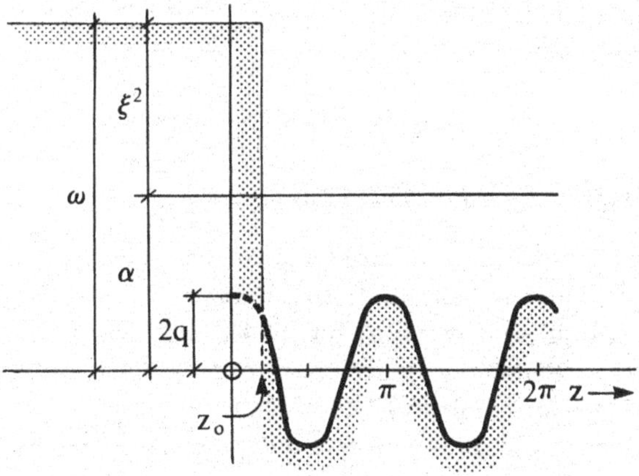

Fig. 6.1. Surface region of a crystal with sinusoidal potential. (After Levine 1968.)

The potential profile in the surface region is shown in Fig. 6.1, where the crystal is terminated at an arbitrary value of $z = z_0$. Inside the crystal, which is of infinite extent for $z \geq z_0$, the potential is sinusoidal, and the surface-state wave function satisfying Mathieu's equation (2.192) has the form

$$\psi_i(z,q) = \phi(z,q)e^{-\mu(z-z_0)}, \ \mu > 0 \tag{6.1}$$

so that

$$\lim_{z\to\infty} \psi_i(z,q) = 0 \tag{6.2}$$

α being the corresponding eigenvalue.

In the vacuum region, beyond the crystal surface ($z \leq z_0$), a constant potential V_0 exists, which in the dimensionless units of (2.193) translates into

$$\omega = V_0/T \ . \tag{6.3}$$

The vacuum wave function has the form

$$\psi_0(z) = Ae^{-\xi(z_0-z)} \tag{6.4}$$

A being a constant, and

$$\xi^2 = \omega - \alpha = E_0/T \tag{6.5}$$

where E_0 is the energy relative to the vacuum level. $\psi_0(z)$ has the correct asymptotic behaviour, namely,

$$\lim_{z\to-\infty} \psi_0(z) = 0 \ . \tag{6.6}$$

6.2 Single-Gap States

It is now necessary to find an explicit form for $\phi(z,q)$ in (6.1). The most general form is a linear combination (or *hybrid*) of all the bulk standing wave states $se_r(z,q)$ and $ce_r(z,q)$ for all FEG r, which were discussed in §2.5. However, a very good approximate expression is obtained for $\phi(z,q)$ by noting that, within the *mth* FEG, the surface state is largely a hybrid of the $ce_m(z,q)$ and $se_m(z,q)$ functions bordering the FEG (Fig. 2.14). Thus,

$$\phi(z,q) = \lambda ce_m(z,q) - se_m(z,q) \tag{6.7}$$

where λ is the *hybridization parameter*. The above approximation holds if the FEG width is narrow compared to the bandwidths, i.e., $q < \alpha/2$, as represented by the broken line in Fig. 2.14(a).

Hence, from (6.1) and (6.7), the unnormalized form of the surface-state wave function is

$$\psi_i(z,q) = [\lambda ce_m(z,q) - se_m(z,q)]e^{-\mu z} \ . \tag{6.8}$$

Since $se_m(z,q)$ and $ce_m(z,q)$ are odd and even functions, respectively, replacing z by $-z$ in (6.8) gives

$$\psi_i(-z,q) = [\lambda ce_m(z,q) + se_m(z,q)]e^{\mu z} \tag{6.9}$$

which *increases* exponentially as $z \to \infty$, and is therefore *rejected* as a surface-state solution for a semi-infinite crystal in the positive half-plane, even though it satisfies Mathieu's equation (2.192), viz.,

$$\psi_i''(z,q) + (\alpha \ - \ 2q\cos 2z)\psi_i(z,q) = 0 \ . \tag{6.10}$$

Inserting (6.8) in (6.10) yields

$$\begin{aligned} &\lambda[ce_m''(z,q) - 2\mu ce_m'(z,q) + (\alpha + \mu^2 - 2q\cos 2z)ce_m(z,q)] \\ &-[se_m''(z,q) - 2\mu se_m'(z,q) + (\alpha + \mu^2 - 2q\cos 2z)se_m(z,q)] = 0 \ . \end{aligned} \tag{6.11}$$

However, $ce_m(z,q)$ and $se_m(z,q)$ are also eigenfunctions of the Mathieu equation with eigenvalues a_m and b_m, so

$$\begin{aligned} ce_m''(z,q) \quad &+ \quad (a_m - 2q\cos 2z)ce_m(z,q) = 0 \\ se_m''(z,q) \quad &+ \quad (b_m - 2q\cos 2z)se_m(z,q) = 0 \ . \end{aligned} \tag{6.12}$$

Eliminating the second derivatives between (6.11) and (6.12) gives

$$\begin{aligned} &\lambda[(\alpha + \mu^2 - a_m)ce_m(z,q) - 2\mu ce_m'(z,q)] \\ &- [(\alpha + \mu^2 - b_m)se_m(z,q) - 2\mu se_m'(z,q)] = 0 \ . \end{aligned} \tag{6.13}$$

On multiplying by $ce_m(z,q)$ and integrating over $0 < z < 2\pi$, and then repeating the process with $se_m(z,q)$, leads to the two equations

$$\begin{aligned} (\alpha + \mu^2 - a_m)\lambda + 2\mu m \quad &= \quad 0 \\ -2\mu m\lambda + (\alpha + \mu^2 - b_m) \quad &= \quad 0 \end{aligned} \tag{6.14}$$

where use has been made of the relations (Jahnke and Emde 1945, McLachlan 1951, Levine 1968)

$$\begin{aligned}
\textstyle\int_0^{2\pi} ce_m^2(z,q)dz = \int_0^{2\pi} se_m^2(z,q)dz &= \pi \\
\textstyle\int_0^{2\pi} ce_m(z,q)se_m(z,q)dz &= 0 \\
\textstyle\int_0^{2\pi} ce_m(z,q)ce_m'(z,q)dz &= 0 \\
\textstyle\int_0^{2\pi} se_m(z,q)se_m'(z,q)dz &= 0 \\
\textstyle\int_0^{2\pi} ce_m(z,q)se_m'(z,q)dz &= K_{mm} \sim m\pi \\
\textstyle\int_0^{2\pi} se_m(z,q)ce_m'(z,q)dz &= -K_{mm} \sim -m\pi \ .
\end{aligned} \tag{6.15}$$

Eliminating λ in (6.14) gives the quadratic equation in μ^2

$$\mu^4 + \mu^2(4m^2 + 2\alpha - a_m - b_m) + (\alpha - a_m)(\alpha - b_m) = 0 \ . \tag{6.16}$$

For an energy lying in a *narrow* FEG (i.e. $b_m \leq \alpha \leq a_m$),

$$\begin{aligned}
&\mu^4 << 1 \ , && 4m^2 >> 2\alpha - a_m - b_m \\
&4m^2 >> \alpha - b_m \ , && 4m^2 >> a_m - \alpha
\end{aligned} \tag{6.17}$$

in which case (6.16) reduces to

$$4m^2\mu^2 + (\alpha - a_m)(\alpha - b_m) = 0 \tag{6.18}$$

or

$$\mu = [(a_m - \alpha)(\alpha - b_m)]^{1/2}/2m \tag{6.19}$$

where the positive root is taken, because $\mu > 0$ in the definition (6.1) of a surface state.

In view of (6.17), equations (6.14) may be written in matrix form, i.e.,

$$\begin{bmatrix} \alpha - a_m & 2\mu m \\ -2\mu m & \alpha - b_m \end{bmatrix} \begin{bmatrix} \lambda \\ 1 \end{bmatrix} = 0 \ . \tag{6.20}$$

The secular equation of (6.20) yields the energies

$$\alpha_{\pm} = \{(a_m + b_m) \pm [(a_m - b_m)^2 - 16m^2\mu^2]^{1/2}\}/2 \tag{6.21}$$

which shows that if $\mu = 0$ (i.e., no surface and hence no surface state), then

$$\alpha_+ = a_m \ , \quad \alpha_- = b_m \tag{6.22}$$

are the band-edge energies. Consequently, the presence of a surface allows $\mu > 0$, which acts as a surface 'perturbation' and, in so doing, *attracts* α_+ and α_- towards each other and the FEG centre. This attraction arises because of the minus sign in the discriminant of (6.21), which in turn is due to the antisymmetric matrix in (6.20).

Neglecting μ^2 in (6.14) shows that

$$\lambda = 2\mu m/(a_m - \alpha) = (\alpha - b_m)/2\mu m \ . \tag{6.23}$$

Thus, since $\mu > 0$ and $b_m \leq \alpha \leq a_m$, $\lambda > 0$, which may be regarded as a 'surface-state rule'. Another expression for λ is

$$\lambda = [(\alpha - b_m)/(a_m - \alpha)]^{1/2} \tag{6.24}$$

which follows from (6.19) and (6.23). Rearranging (6.19) and (6.24) enables α and μ to be written as single-valued functions of λ, namely,

$$\alpha = (b_m + \lambda^2 a_m)/(1 + \lambda^2) \tag{6.25}$$

$$\mu = [\lambda/(1 + \lambda^2)][(a_m - b_m)/2m] \ . \tag{6.26}$$

When $\lambda = 0$, $\alpha = b_m$ and $\mu = 0$, there is no surface state, and the energy lies at the bottom of the FEG. For $\lambda = 1$,

$$\alpha = (a_m + b_m)/2 \ \text{and} \ \mu = (a_m - b_m)/4m \ .$$

Here, a surface state exists whose energy is in the middle of the FEG, and the bulk *damping (or localization) parameter* μ is at its maximum. When $\lambda = \infty$, $\alpha = a_m$ and $\mu = 0$, so again no surface state occurs, and the energy coincides with the top of the FEG.

In closing this section, it is worthwhile to examine the case of $\lambda < 0$, which implies that $\mu < 0$ from (6.23) and (6.26). Such a situation corresponds precisely to the one encountered in (6.9), and is rejected as a physically unreal surface-state solution. Finally, as can be seen from (6.24), λ is pure imaginary when α lies in the allowed bands, i.e., $b_m > \alpha > a_m$. μ is then also pure imaginary by (6.23) or (6.26), and (6.8) becomes a Bloch wave of a bulk-state solution.

6.3 Continuity Condition

The condition governing the existence of a surface state is derived from the continuity of the wave function and its derivative at the surface $z = z_0$, i.e.,

$$\begin{aligned} \psi_0(z_0) &= \psi_i(z_0, q) \\ \psi_0'(z_0) &= \psi_i'(z_0, q) \,. \end{aligned} \tag{6.27}$$

Substituting (6.1) and (6.4) in (6.27) gives

$$\phi(z_0, q)/\phi'(z_0, q) = (\xi + \mu)^{-1} \tag{6.28}$$

where, for many physical situations, the right-hand side is small enough to be treated as a perturbation. In particular, for the extreme case of an infinitely high step potential, where $\xi \to \infty$, (6.28) reduces to the simple nodal condition $\phi(z_0, q) = 0$. Now (6.7) in (6.28) leads to the general continuity condition

$$\lambda = \frac{se_m(z_0, q) - (\xi + \mu)^{-1} se_m'(z_0, q)}{ce_m(z_0, q) - (\xi + \mu)^{-1} ce_m'(z_0, q)} > 0 \tag{6.29}$$

which must be satisfied for a surface state to occur.

Although (6.29) is a rather complicated expression, its analysis may be simplified by discussing certain special cases, which require a knowledge of the Mathieu functions at their symmetry points, such as $z_0 = 0$ and $\pi/2$. Expanding $se_m(z, q)$ and $ce_m(z, q)$ as Fourier cosine and sine series, i.e.,

Eigenfunction	Eigenvalue
$ce_{2n}(z, q) = \sum_{r=0}^{\infty} A_{2r}^{(2n)} \cos 2rz$	a_{2n}
$ce_{2n+1}(z, q) = \sum_{r=0}^{\infty} A_{2r+1}^{(2n+1)} \cos(2r + 1)z$	a_{2n+1}
$se_{2n+1}(z, q) = \sum_{r=0}^{\infty} B_{2r+1}^{(2n+1)} \sin(2r + 1)z$	b_{2n+1}
$se_{2n+2}(z, q) = \sum_{r=0}^{\infty} B_{2r+2}^{(2n+2)} \sin(2r + 2)z$	b_{2n+2}

(6.30)

shows that the symmetries of these functions fall naturally into four classes, depending on m being even or odd. Using (6.30), Table 6.1 may be compiled, which lists the values of the Mathieu functions and their first derivatives with respect to z at $z = 0$ and $\pi/2$. The numbers N_i in Table 6.1 are positive by definition, and are functions of q, because A and B in (6.30) are functions of q (McLachlan 1951).

Function	$z = 0$		$z = \pi/2$	
	m = even	m = odd	m = even	m = odd
$ce_m(z,q)$	N_1	N_1	$(-1)^{m/2}N_2$	0
$se_m(z,q)$	0	0	0	$(-1)^{(m-1)/2}N_3$
$ce'_m(z,q)$	0	0	0	$-m(-1)^{(m-1)/2}N_4$
$se'_m(z,q)$	mN_5	mN_5	$m(-1)^{m/2}N_6$	0

Table 6.1 Values of Mathieu functions at symmetry points.

As the first special example, consider the situation where $z_0 = 0$, which corresponds to the termination of the crystal at a potential *maximum* in the unit cell. Here, surface states of the Shockley type (§4.3 and §5.2) may be expected to occur, if they exist for a sinusoidal potential. According to Table 6.1, for $z_0 = 0$, (6.29) becomes

$$\lambda = -mN_5/N_1(\xi + \mu) \tag{6.31}$$

m being odd or even. Thus, since $N_i > 0$, it follows that $\lambda < 0$, so *no* Shockley surface state arises for a sinusoidal potential.

When $z_0 = \pi/2$, the crystal is terminated at a potential *minimum*, so a surface state of the Tamm type (§4.1 and §5.2) may appear. Such an asymmetric termination of the potential might distort the rigid covalent bond between the atoms, so Tamm states are also referred to as *covalent* surface states (Levine 1968). For $z_0 = \pi/2$, equation (6.29) gives

$$\lambda = -mN_6/N_2(\xi + \mu), \quad m \text{ even} \tag{6.32}$$

$$\lambda = (\xi + \mu)N_3/mN_4, \quad m \text{ odd} . \tag{6.33}$$

Hence, a covalent surface state will (will not) exist for m odd (even), i.e., at the edge X (centre Γ) of the one-dimensional Brillouin zone (Fig. 2.14b).

As ξ/m increases, λ increases, and by (6.25) the surface state moves nearer to the top of the FEG, until it disappears into the top of the FEG when $\lambda = \infty$, which corresponds to the case of an infinite step potential, $\xi = \infty$.

In order to discuss an *arbitrary termination* z_0, the simplest model is considered, namely, that of a nearly-empty lattice ($q \sim 0$) and infinitely high surface-potential barrier ($\xi \sim \infty$). Since

$$ce_m(z_0, 0) = \cos mz_0$$

and (6.34)

$$se_m(z_0, 0) = \sin mz_0$$

by (2.195) and (2.203), equation (6.29) reduces to

$$\lambda \simeq \tan mz_0 \tag{6.35}$$

so $\lambda > 0$ only when

$$\begin{aligned} 0 &< mz_0 < \pi/2 \\ \pi &< mz_0 < 3\pi/2 \\ 2\pi &< mz_0 < 5\pi/2 \ldots \end{aligned} \tag{6.36}$$

which are the surface-state existence conditions. From (6.25), (6.26) and (6.35), the energies α and damping constant μ are

$$\alpha = b_m \cos^2 mz_0 + a_m \sin^2 mz_0 \tag{6.37}$$

$$\mu = [(a_m - b_m)/4m] \sin 2mz_0 \; . \tag{6.38}$$

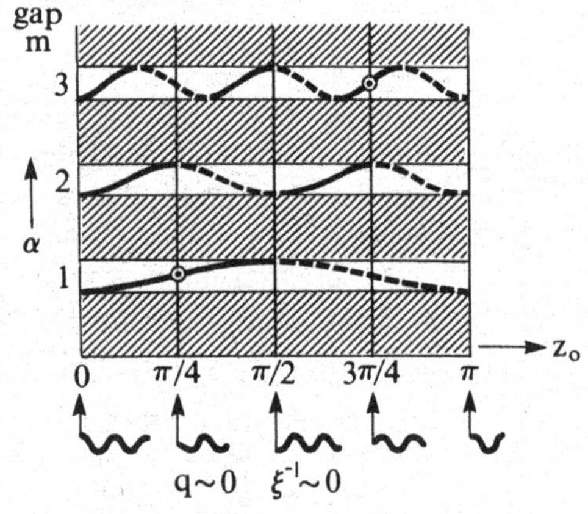

Fig. 6.2. Surface-state energies versus z_0 for nearly-empty lattice and infinite surface potential. (After Levine 1968.)

A plot of α versus z_0 is shown in Fig. 6.2 for $0 < z_0 < \pi$, where the surface states are depicted by the sinusoidal curves in the various FEG, and crystal terminations at the bottom of the figure. In the first ($m = 1$) FEG, equation (6.36) permits a surface state, such as that represented by the open circle at $z_0 = \pi/4$, to exist in the range $0 < z_0 < \pi/2$. As can be seen, the energy increases smoothly from the bottom to the top of the FEG as z_0 goes from 0 to $\pi/2$. However, for $z_0 > \pi/2$, $\lambda < 0$, so an *unallowed* surface state occurs, as indicated by the broken line. This state has the correct α and continuity condition of the wave function, but the *incorrect* behaviour at $z \to \infty$. In the second ($m = 2$) FEG, surface states appear in the two regions $0 < z_0 < \pi/4$ and $\pi/2 < z_0 < 3\pi/4$. In general, the mth FEG has m different surface-state regions. At $z_0 = 0$, all energies α lie at the bottom band edges b_1, b_2, ..., while at $z_0 = \pi/2$ the energies α alternate a_1, b_2, a_3, b_4, ...

Consider now the more general situation of $q \sim 0$ and ξ being finite, which corresponds to a nearly-empty lattice with a finite step potential. Here, (6.34) in (6.29) gives

$$\lambda = \frac{\sin mz_0 \ - \ m(\xi \ + \ \mu)^{-1} \cos mz_0}{\cos mz_0 \ + \ m(\xi \ + \ \mu)^{-1} \sin mz_0} \ . \tag{6.39}$$

For a large finite barrier, $\xi >> \mu$ and $m\xi^{-1} << 1$, so $\sin m\xi^{-1} \simeq m\xi^{-1}$ and $\cos m\xi^{-1} \simeq 1$. Hence, (6.39) may be written as

$$\lambda \simeq \tan m(z_0 \ - \ \xi^{-1}) \ . \tag{6.40}$$

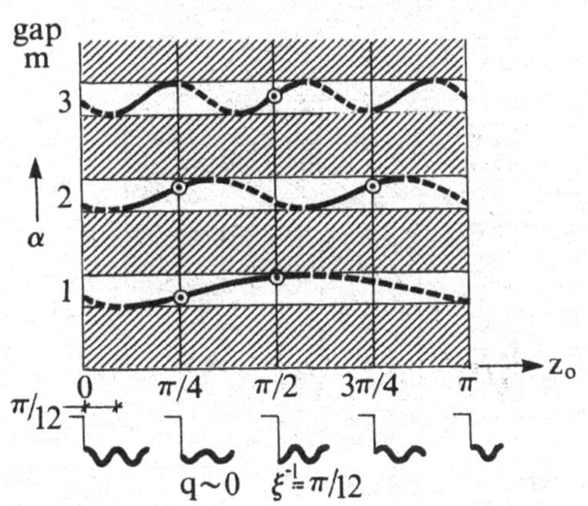

Fig. 6.3. Surface-state energies versus z_0 for nearly-empty lattice and finite surface potential. (After Levine 1968.)

Equation (6.40) differs from (6.35) by the phase-shift term $m\xi^{-1}$, which may be regarded as a small perturbation on the previous results, and corresponds to a thin slab of thickness $z = \xi^{-1}$ being removed from the crystal.

The surface-state expressions corresponding to (6.36) and (6.37) are obtained by merely replacing z_0 by $(z_0 - \xi^{-1})$, so the α vs. z_0 diagram can be obtained immediately by translating Fig. 6.2 uniformly to the right by an amount ξ^{-1}. Figure 6.3 shows the result of this operation for $\xi^{-1} = \pi/12$. It should, of course, be noted that the bands higher than $m = 2$ are not accurately represented, since the inequality $m\xi^{-1} << 1$ is no longer valid. As shown by the open circles, covalent surface states now appear at $z_0 = \pi/2$ in all $m =$ odd FEG, in agreement with the previous calculations. Likewise, no Shockley states are present at $z_0 = 0$. The surface states at $z_0 = \pi/4$ and $3\pi/4$ in the $m = 2$ FEG are called *ionic* surface states (Levine 1968), because, as will be shown later, they can be identified with the surface state of an MX-type ionic crystal.

The next stage in the generalization of the analysis is to investigate a crystal with $q > 0$, and initially take $\xi \to \infty$. Such a crystal has an infinite surface potential and a strong periodic potential, so the NFE method is no longer applicable. Under these circumstances, (6.29) becomes

$$\lambda = se_m(z_0, q)/ce_m(z_0, q) \ . \tag{6.41}$$

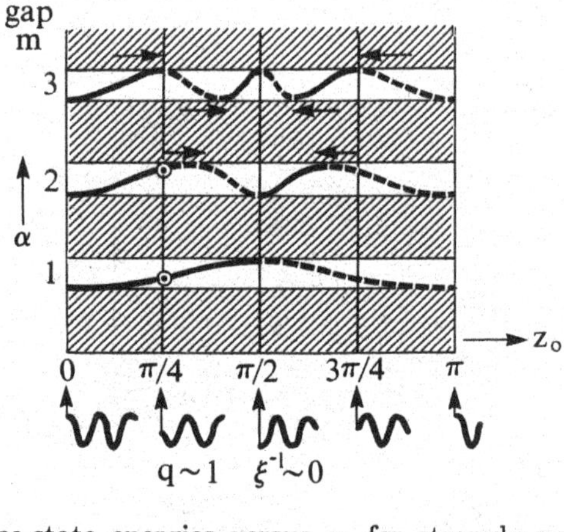

Fig. 6.4. Surface-state energies versus z_0 for strongly periodic lattice and infinite surface potential. Note compression towards $\pi/2$. (After Levine 1968.)

The effect of increasing q on Mathieu functions has already been discussed in conjunction with Fig. 2.13. There it was demonstrated that, as q increases, the amplitudes and nodes of $ce_m(z,q)$ and $se_m(z,q)$ are drawn closer to $z = \pi/2$, where the attractive potential is greatest. Consequently, the z_0-points, at which $\lambda = 0$ or ∞, are pulled closer and closer to $z_0 = \pi/2$ as q increases. The shift of the nodes is represented by the arrows in Fig. 6.4, which may be obtained by simply 'compressing' Fig. 6.2.

Finally, in the most general situation, $q > 0$ and ξ is finite, so (6.30) in (6.29) gives

$$\lambda = \frac{\sum_{r=0}^{\infty} B_{2r+1}^{(2n+1)}\ [\sin(2r+1)z_0\ -\ (2r+1)(\xi+\mu)^{-1}\cos(2r+1)z_0]}{\sum_{r=0}^{\infty} A_{2r+1}^{(2n+1)}\ [\cos(2r+1)z_0\ +\ (2r+1)(\xi+\mu)^{-1}\sin(2r+1)z_0]} \tag{6.42}$$

for m odd. Taking ξ to be sufficiently large, so that $\xi >> \mu$ and $(2r+1)\xi^{-1} << 1$ for all r of interest, then

$$\lambda = se_{2n+1}[(z_0 - \xi^{-1}), q]/ce_{2n+1}[(z_0 - \xi^{-1}), q] \tag{6.43}$$

is obtained from (6.42), in the same way as (6.40) was from (6.39). Similar equations can also be found for m even. The structure of (6.43) shows that it contains both the shift and compression aspects of the earlier cases. Thus, Fig. 6.5 illustrates the result of *superposing* the effects of shifting

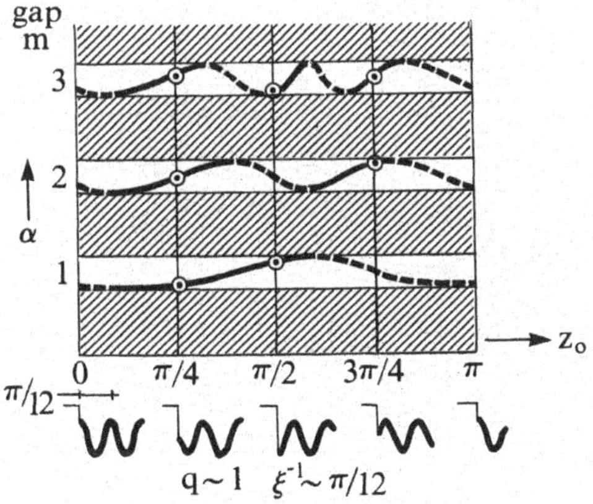

Fig. 6.5. Surface-state energies versus z_0 for strongly periodic lattice and finite surface potential. Covalent states at m = odd and $z_0 = \pi/2$. Ionic states at $m = 2$ and $z_0 = \pi/4, 3\pi/4$. (After Levine 1968.)

Fig. 6.2 to the right by an amount $\xi^{-1} = \pi/12$ and compressing it by an amount proportional to $q = 1$. Note that there are still no Shockley states at $z_0 = 0$, and that covalent surface states still appear at $z_0 = \pi/2$ for m odd. The absence of Shockley surface states throughout the above analysis is not really surprising, because these states also require bulk band-crossing (Statz 1950), which does not occur for the Mathieu potential (Fig. 2.14a).

Comparing Figs. 6.3 and 6.5, it is clear that, in the FEG $m = 2$, the so-called ionic surface state at $z_0 = \pi/4 (z_0 = 3\pi/4)$ has moved nearer to the *lower* (*upper*) band edge. This effect increases with q, which is a measure of the ionicity. Thus, in accord with the semi-classical theory of Levine and Mark (1966) for MX-type ionic crystals, one surface state is M-like and the other X-like, corresponding to $z_0 = 3\pi/4$ and $\pi/4$ lattice terminations, respectively. These findings are also consistent with those of Levine and Davison (1968), who modelled an ionic crystal by assigning s- and p-orbitals to electrons on alternate M- and X-sites along a one-dimensional chain of atoms.

7

FIELD, VIRTUAL AND IMAGE STATES

Between the idea and the reality
Between the motion and the act
Falls the Shadow.
— Thomas S. Eliot

So far, it has been assumed that the potential outside the crystal is *constant*, i.e., $V(x) = V_0$ in the vacuum region (the so-called *abrupt-potential model*). One should remember, however, that this is only an assumption, which greatly simplifies the direct-matching procedure, and enables surface-state calculations to be performed in a straightforward manner.

In reality, an electron outside the crystal can experience different potential fields, depending on the physical conditions to which the crystal surface is exposed. We shall consider two of the most interesting and important cases, viz., when an external electric field is applied to a crystal and when the image potential is taken into account, i.e., the potential generated by the positive image charge created when an electron leaves the crystal.

7.1 Effect of External Electric Field

Let us assume that the crystal surface is located at $x = 0$ and that the external electric field, applied to a crystal, is linear. If the vacuum region, with a constant surface potential $V(x) = V_0$, extends to the right of the crystal, i.e., for $x > 0$, then

$$V(x) = V_0 \pm Fx \tag{7.1}$$

where F is the electric field intensity and $+(-)$ sign corresponds to the positive (negative) field (Fig. 7.1).

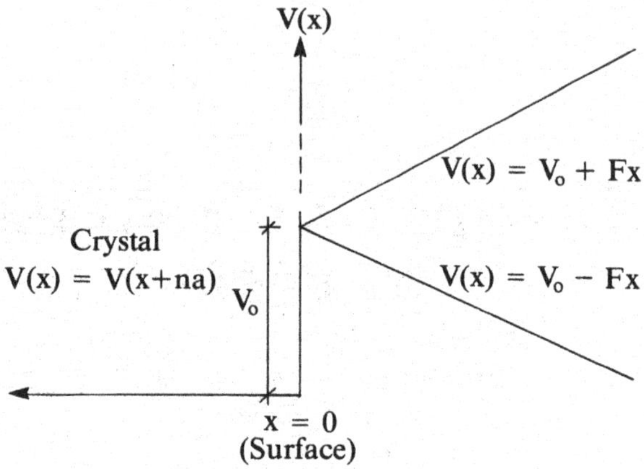

Fig. 7.1. Diagram showing model of potential profile in applied electric field situations. F is gradient of field and V_0 the vacuum level.

The Schrödinger equation may now be written as

$$\psi''(x) + [E - (V_0 \pm Fx)]\psi(x) = 0 \tag{7.2}$$

which on setting

$$z = (-E + V_0 \pm Fx)F^{-2/3} \tag{7.3}$$

becomes the well-known *Airy equation* (Jeffreys and Jeffreys 1962, Landau and Lifshitz 1965)

$$\psi''(z) - z\psi(z) = 0\ . \tag{7.4}$$

The general solution of (7.4) is a linear combination of Airy functions, namely,

$$\psi(z) = c_1 Ai(z) + c_2 Bi(z) \tag{7.5}$$

c_1 and c_2 being constants. A plot of the Airy functions is shown in Fig. 7.2.

To find the energy spectrum of the surface states, it is also necessary to know the form of the wave function inside the crystal, i.e., for $x < 0$. Assuming that the crystal is represented by a δ-barrier KP model, we have, for $-a < x < 0$,

$$\psi_c(x) = c_3[e^{ik_1(x+a)} + \lambda e^{-ik_1(x+a)}]e^{\mu a} \tag{7.6}$$

via (2.53), (2.9) and (3.7), where λ is given by (3.13b). Let us now consider two cases.

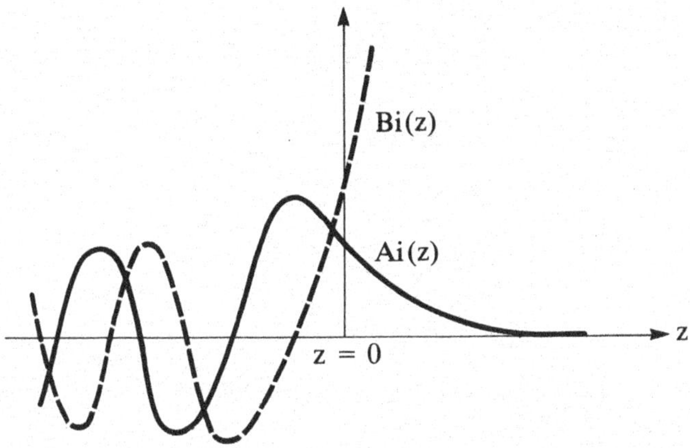

Fig. 7.2. Plot of Airy functions $Ai(z)$ (solid curve) and $Bi(z)$ (broken curve).

7.1.1 Negative field

Here $V(x) = V_0 - Fx$ and (7.3) becomes $z = (-E + V_0 - Fx)F^{-2/3}$, so as $x \to +\infty$, $z \to -\infty$. However, for negative values of z, both Airy functions $Ai(z)$ and $Bi(z)$ always remain finite (Fig. 7.2), so, the wave function in the vacuum region must have the form of (7.5). Consequently, the requirements $\psi_c(0) = \psi(z_0)$ and $\psi_c'(0) = \psi'(z_0)$, where $z_0 = z(x = 0)$, lead to two equations for three unknowns. Symbolically, they can be written as

$$\begin{aligned} F_1(c_1, c_2, c_3) &= 0 \\ F_2(c_1, c_2, c_3) &= 0\,. \end{aligned} \tag{7.7}$$

In physical language, (7.7) means that, when a negative electric field is applied to a crystal, electrons with energies $E < V_0$ can *tunnel* into the vacuum through the potential barrier, as in field-emission microscopes. Under such conditions, localized surface states become *virtual* surface states, whose basic properties will be described in §7.1.3 and §8.3.2.

7.1.2 Positive field

In this case, $V(x) = V_0 + Fx$ and (7.3) takes the form $z = (-E + V_0 + Fx)F^{-2/3}$, which, for $x \to +\infty$, gives $z \to +\infty$. Since the only solution that remains finite for all positive values of z is $Ai(z)$ (Fig. 7.2), $c_2 = 0$ in (7.5), so

$$\psi(z) = c_1 Ai(z) \tag{7.8}$$

and the matching conditions (7.7) become

$$\begin{aligned} G_1(c_1, c_3) &= 0 \\ G_2(c_1, c_3) &= 0 . \end{aligned} \tag{7.9}$$

Equations (7.9) have non-trivial solutions for such energies E that make the determinant of the system vanish.

The wave functions corresponding to these energy levels, (7.6) and (7.8), are damped away from the surface in the directions of the vacuum and the crystal. Thus, electrons occupying these levels are truly localized at the crystal surface.

An interesting possibility arises here and provides a simplified model of the field-ion microscope (Müller and Tsong 1969, 1973). When no external field is applied, the usual requirement on the electron energy is $E < V_0$. However, when a positive electric field is applied, electrons with $E > V_0$ still cannot leave the crystal (Fig. 7.1), since the wave function (7.8) provides an effective localization of an electron in the vicinity of a crystal surface (Fig. 7.2). Surface states with $E > V_0$ were introduced as field-sustained states (Stęślicka 1976, Thakkar and Stęślicka 1978), and are referred to here as simply *field* states.

(i) *Field states*

Applying the standard matching conditions to (7.6) and (7.8) gives

$$e^{-\mu a} = \cos\xi - \delta\xi^{-1}\sin\xi \tag{7.10}$$

where

$$\delta = aF^{1/3}Ai'(z_0)/Ai(z_0) \tag{7.11}$$

and

$$z_0 \equiv z(x = 0) = (V_0 - E)F^{-2/3} \tag{7.12}$$

from (7.3). Combining (7.10) and (3.17) leads to the surface-state energy expression

$$\xi \cot \xi = [\xi^2 + \delta(\delta + 2p)]/2p \tag{7.13a}$$

with

$$p + \delta > 0\,. \tag{7.13b}$$

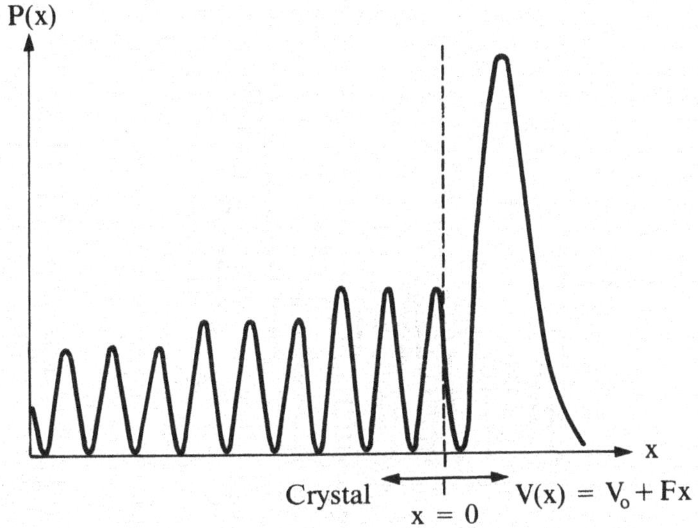

Fig. 7.3. Relative probability density function $P(x)$ for third FEG ($E > V_0$).

Numerical solution of (7.13) shows that the effect of the applied field on the normal surface states ($E < V_0$) is to raise their energy without changing the form of the relative probability density function $P(x)$ of (3.22), shown in Figs. 7.3 and 7.4 for the third and fourth FEG, respectively, for which $E > V_0$. The behaviour of $P(x)$ inside the crystal is nearly identical to that for normal surface states (cf. Fig. 3.5). However, *outside* the crystal, it is considerably different, since $P(x)$ for field states displays a local maximum (Fig. 7.3) or a number of local maxima (Fig. 7.4). Moreover, the amplitudes of these maxima grow with the distance from the surface, until the global maximum is attained; thereafter $P(x)$ decreases monotonically. For example, $P(x)$ for the field state in the fifth FEG has as many as nine local maxima outside the crystal. Thus, the presence of an external positive electric field has interesting consequences for the localization properties of electrons in surface states. As shown in Chapter 3, electrons in surface states in the first FEG ($E < V_0$) tend to be strongly localized near the surface and *inside* the crystal (Fig. 3.4). Electrons in field states tend to be strongly localized *outside* the crystal at distances which may be as large as a few lattice constants. Consequently, one expects field states to be involved in the field-ionization process (Stęślicka 1977, Stęślicka and Perkal 1977, Thakkar and Stęślicka 1978).

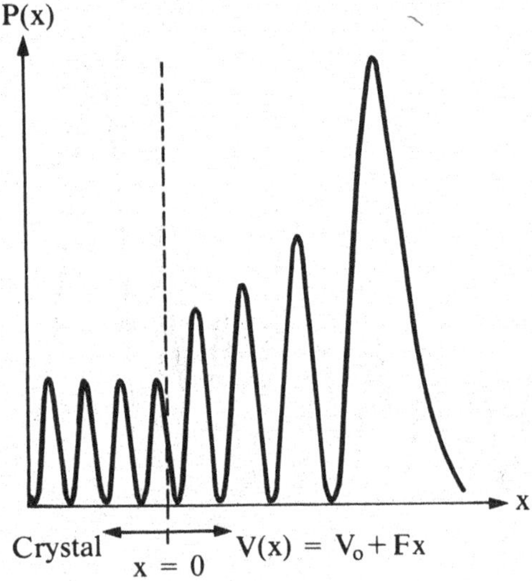

Fig. 7.4. $P(x)$ for fourth FEG ($E > V_0$).

It can be shown that the Tamm model (Chapter 3) is the zero-field limit of the positive field model. When $F \to 0$, $z_0 \to \infty$, by (7.12) for $E < V_0$. However, for large values of z_0, the asymptotic expansions of the Airy functions can be used, namely, (Abramowitz and Stegun 1972)

$$\begin{aligned} Ai\,(z_0) &= \tfrac{1}{2}\, z_0^{-1/4} e^{-S} \\ Ai'(z_0) &= -\tfrac{1}{2}\, z_0^{1/4}\, e^{-S} \\ S &= \tfrac{2}{3}\, z_0^{3/2}\,. \end{aligned} \tag{7.14}$$

Inserting (7.14) in (7.11) gives

$$\delta = -a z_0^{1/2}\, F^{1/3} \tag{7.15}$$

which, via (7.12), (3.15) and (3.6), becomes

$$\delta = -(q^2 - \xi^2)^{1/2} . \tag{7.16}$$

Thus, (7.10) and (7.13) coincide with (3.14) and (3.18), respectively.

7.1.3 Virtual surface states of δ-barrier crystal

As shown in §7.1.1, any attempt to study the surface-state behaviour, when an external negative field is applied, leads to a set of equations of the form (7.7). To solve these equations, it is necessary either to introduce an additional boundary condition (when the direct-matching procedure is used), or to develop a method of calculation which will avoid the above difficulties. The latter was achieved by Davison *et al.* (1980), who employed the Green-function factorization method of Garcia-Moliner and Rubio (1969) discussed in §8.3. Here, the direct-matching procedure is used to treat this problem.

The concept of *virtual surface states*, in the applied negative electric field situation, was introduced by Modinos (1974). The idea is similar to that of a virtual atomic-like state originally proposed by Anderson (1961). Heine (1965) also observed that, at a semiconductor-metal interface, a true surface state cannot exist, because the wave function in the metal contains travelling waves as well as decaying ones.

The effect of an applied linear field on the surface states of a δ-barrier KP crystal was examined by Davison and Tan (1971). As an additional condition, enabling (7.7) to be solved, they introduced a 'horizontal termination' of the potential $V(x)$ in the vacuum region ($x \in (0, -\infty)$), i.e., $V(x) = p\sum_{n=1}^{\infty} \delta(x - na)$ for $x \geq 0$, $V(x) = V_0 + Fx$ for $-x_1 \leq x < 0$, $V(x) = V_1$ (constant) for $-\infty < x < -x_1$. As a result, the restriction on the electron energy $E < V_1$ was introduced and the tunnelling situation ($E > V_1$), leading to the formation of a virtual surface state, could not be analysed.

Modinos (1974) addressed this situation by extending the linear field region to occupy the whole negative half-space ($x \in (0, -\infty)$). An extra relation between c_1, c_2 and c_3 in (7.7) was obtained from the normalization condition for the wave function in the linear-field region. The result was that, on increasing F, the discrete surface-state energy level moved to lower energy values and broadened into a virtual-surface-state peak.

(i) *Boundary condition and energy expression*

Stęślicka and Perkal (1980) used the idea of 'box normalization' to introduce the required additional boundary condition, i.e.,

$$\psi(x = d) = 0, \quad d >> a \,. \tag{7.17}$$

Applying standard matching conditions to (7.5) and (7.6), imposing (7.17) on (7.5) and taking (3.17) into account leads to the equation for the surface-state energy, viz.,

$$\xi \cot\xi = [\xi^2 + \Delta(\Delta - 2p)]/2p \tag{7.18a}$$

$$p - \Delta > 0 \tag{7.18b}$$

where (cf. 7.11)

$$\Delta = aF^{1/3} \frac{Ai(z_1)Bi'(z_0) - Ai'(z_0)Bi(z_1)}{Ai(z_1)Bi\,(z_0) - Ai\,(z_0)Bi(z_1)} \tag{7.19}$$

with (cf. 7.12)

$$z_1 \equiv z(x = d) = F^{-2/3}(-E + V_0 - Fd) = z_0 - F^{1/3}d \,. \tag{7.20}$$

Assuming $F \sim 0.1\ V/\text{Å}$ and d of the order of a few centimetres gives $z_0 \sim 10$ and $z_1 \sim -10^7$ by (7.20). However, for large negative values of z_1, the following asymptotic expansions of the Airy functions may be used (Abramowitz and Stegun 1972), viz.,

$$\begin{aligned} Ai(z_1) &= \pi^{-1/2} \mid z_1 \mid^{-1/4} \sin\beta \\ Bi(z_1) &= \pi^{-1/2} \mid z_1 \mid^{-1/4} \cos\beta \end{aligned} \tag{7.21}$$

where

$$\beta = \tfrac{2}{3}|z_1|^{3/2} + \frac{\pi}{4}. \tag{7.22}$$

Substituting (7.21) in (7.19) yields

$$\Delta = aF^{1/3} \frac{\tan\beta\ Bi'(z_0) - Ai'(z_0)}{\tan\beta\ Bi\,(z_0) - Ai(z_0)} \,. \tag{7.23}$$

If $|z_1| \sim 10^7$, then $\beta \sim 10^{10}$, and a change in $d \sim 10^{-3}$Å causes a change in $\beta \sim \pi$, which immediately leads to the conclusion that all values of β in the range $(-\ \pi/2, \pi/2)$ are possible. All values of $\tan\beta \in (-\infty, \infty)$ and, via (7.23), of $\Delta \in (-\infty, \infty)$ are therefore also possible. Taking (7.18) into account, restricts Δ to $(-\infty, p)$.

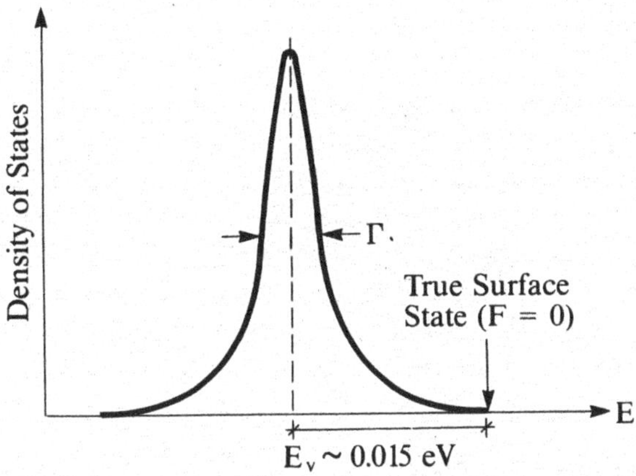

Fig. 7.5. Energy density of states of virtual surface state for $F = 0.1\ V/\text{Å}$ with a lifetime of Γ^{-1}.

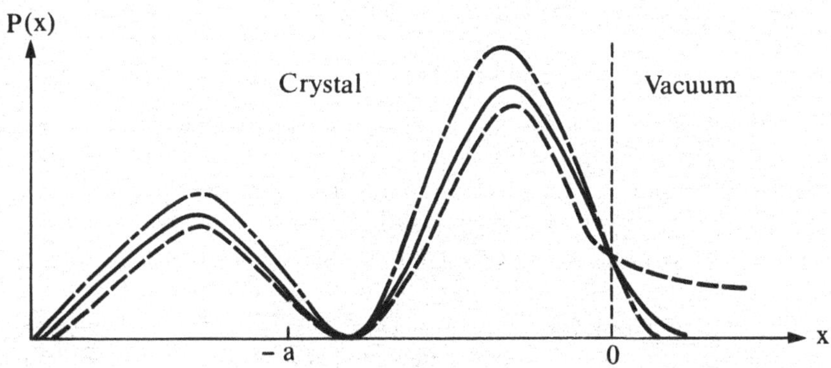

Fig. 7.6. Variation of $P(x)$ in vacuum-crystal region for constant potential (solid line), positive (dashed-dotted line) and negative (dashed line) fields.

Remembering that the first FEG extends from $\xi = 0$ to ξ_1, which is a solution of $\xi_1 \tan \xi_1/2 = p$ (Fig. 2.3), leads to the conclusion that the solutions ξ of (7.18) are spread over the entire FEG region. However, numerical solution of (7.18) indicates that about 80% of the values of ξ lie in the very narrow range $\Delta E \sim 10^{-4}$ eV of the energy. The computed histogram of the density of surface states for $F = 0.1V/\text{Å}$ (Fig. 7.5) shows the energy E_v of the virtual-surface-state peak to be 0.015eV below that of the zero-field surface state, which is in excellent agreement with the Modinos (1974) result.

(ii) *Relative probability density function*

The localization properties of an electron in a virtual surface state can be studied by using the probability density function $P(x)$ (3.22). Figure 7.6 presents $P(x)$ calculated for the energy $E = E_v$ (Fig. 7.5). For comparison purposes, $P(x)$ for a surface state in Tamm's model ($F = 0$) and for a field state (positive external electric field) are also shown. It is worth noting that, outside the crystal, $P(x)$ for a virtual surface state (dashed line) is distinctly different from zero, as one would expect for the tunnelling situation.

7.2 Image States

It is well-known that an electron *outside* a metal surface experiences an electrostatic *image force* (Nordheim 1928). If x is the distance of the electron from the surface, then the *image potential* is

$$V(x) = V_0 - (4x)^{-1} \tag{7.24}$$

where the vacuum region extends from 0 to ∞. Obviously, when $x \rightarrow \infty$, $V(x) \rightarrow V_0$, the vacuum potential level.

Echenique and Pendry (1978) showed that, in a potential field of the form of (7.24), a *Rydberg series* of surface states is generated, which is similar to the hydrogenic states of liquid helium suggested by Cole and Cohen (1969).

The experimental detection of image-potential states on the surfaces of fcc metals (Cu, Ag, Ni), using inverse (Dose *et al.* 1984, Straub and Himpsel 1984, Reihl *et al.* 1987) and two-photon (Giesen *et al.* 1985, 1986, 1987) photoemission aroused new interest in the theory of surface states. In some fcc metals, the FEG in the bulk structure lies above the Fermi level E_f (Fig. 5.6) and the vacuum level ($V_0 = 0$ in Fig. 5.6) is located in the FEG. Surface states appearing in this FEG, and associated with an image potential, should therefore form a Rydberg series of *image states*.

The problem of determining the precise form of the potential barrier V for an electron in close proximity to a metal surface has been discussed in many papers (Rundgren and Malström 1977, Thörner and Borstel 1980,

McRae and Kane 1981, Weinert *et al.* 1985, Read 1986, Jennings *et al.* 1988). The corrugation and dynamical effects in V seem to be relatively unimportant (Jones *et al.* 1984, Barribeau *et al.* 1985), so the potential can be expressed in a one-dimensional form.

7.2.1 Image potential and image plane

The standard form of the image potential with a cut-off is

$$V(x) = V_0 - [4(x - x_0)]^{-1}, \quad x > 0 \tag{7.25}$$

where the positive constant x_0 denotes the *image plane.*

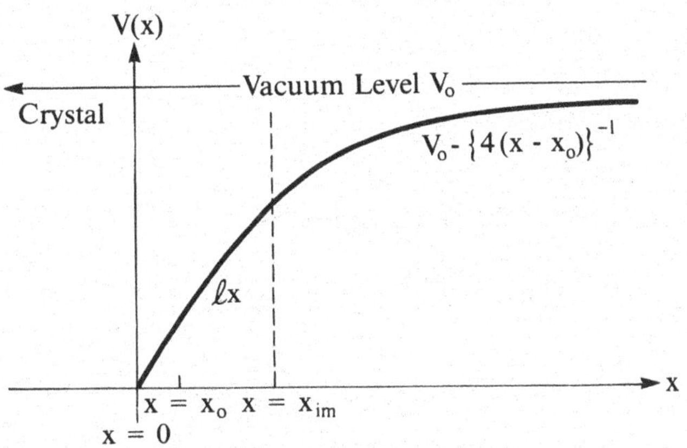

Fig. 7.7. Image potential profile with origin at $x = x_0$ and cut-off at $x = x_{im}$.

The main problem in constructing theoretical potential models of a metal surface is the lack of unique information about the shape of $V(x)$ near the surface, i.e., for $0 < x < x_{im}$, $x_{im} > x_0$ (Fig. 7.7); $x = x_{im}$ is the lower limit of the image force. Such information is usually obtained from the comparison of experimental and theoretical values of surface-state energies, and leads to different values when using different approaches (Thörner and Borstel 1980, McRae and Kane 1981, Read 1986, Jennings *et al.* 1988). Let us assume, therefore, that $V(x)$ is linear for $0 < x < x_{im}$ (Dietz *et al.* 1980, Stęślicka *et al.* 1988), since for this potential the Schrödinger equation reduces to Airy's equation (7.4). Moreover, as shown in §7.2.3, such an approximation gives excellent agreement with experimental data.

Consequently, at a metal surface, it is appropriate to replace (7.25) by (Fig. 7.7)

$$V(x) = \begin{cases} V_\ell = \ell x, & 0 < x < x_{im} \qquad (7.26) \\ V_{im} = V_0 - [4(x - x_0)]^{-1}, & x \geq x_{im} \qquad (7.27) \end{cases}$$

where ℓ is a constant. In addition, we require the image and linear potentials to merge smoothly at $x = x_{im}$, i.e.,

$$\left. \frac{dV_{im}}{dx} \right|_{x=x_{im}} = \ell \,. \tag{7.28}$$

Thus, $x_{im} = f(x_0)$, and we have only one adjustable parameter, namely, that of the image potential plane x_0. To proceed further, it is necessary to solve the Schrödinger equation for (7.26) and (7.27).

7.2.2 Wave functions

(i) *Image potential*

Inserting V_{im} from (7.27) in the Schrödinger equation gives

$$\psi''_{im}(y) + [\alpha^2(E - V_0) + \frac{\alpha}{4y}]\psi_{im}(y) = 0 \tag{7.29}$$

where

$$y = (x - x_0)/\alpha, \qquad x > x_0 > 0 \,. \tag{7.30}$$

Equation (7.29) is a special case of Whittaker's form of the confluent hypergeometric equation, i.e.,

$$\psi''_{im}(y) + [-\tfrac{1}{4} + \frac{\alpha}{y} + \frac{(\frac{1}{4}) - m^2}{y^2}]\psi_{im}(y) = 0 \tag{7.31}$$

(Whittaker 1904, Whittaker and Watson 1958). Indeed, putting

$$m = \tfrac{1}{2}; \quad \alpha^2(E - V_0) = -\tfrac{1}{4} \tag{7.32}$$

converts (7.31) into (7.29).

A fundamental system of solutions of (7.31) is formed by the two Whittaker's functions $W_{\alpha,m}(y)$ and $W_{-\alpha,m}(-y)$, since (7.31) is unaltered if the signs of α and y are changed throughout. For surface-state investigations, $E < V_0$, so α takes only real values by (7.32). However, for α real, y is also real via (7.30). Using the asymptotic expansion of Whittaker's functions (Whittaker and Watson 1958) shows that, for $y > 0$ and large, $W_{-\alpha,1/2}(-y)$ diverges like $y^{-\alpha}\ e^{y/2}$, while $W_{\alpha,1/2}(y)$ has a convergent asymptotic form tending to zero as (Löudon 1959)

$$\lim_{y\to\infty} W_{\alpha,1/2}(y) \sim y^{\alpha}\ e^{-y/2}\ .$$

Since $x > x_0$, it follows from (7.30) that $y > 0$, if $\alpha > 0$, so a suitable wave function is

$$\psi_{im}(x) = c_1 W_{\alpha,1/2}(y), \quad x \geq x_{im} \tag{7.33}$$

c_1 being a constant.

(ii) *Linear potential*

Introducing $V_\ell = \ell x$ (7.26) into the Schrödinger equation leads to Airy's equation (7.4). Thus, the corresponding wave function for $0 \leq x \leq x_{im}$ (Fig. 7.7) has the form

$$\psi_\ell = c_2 Ai(z) + c_3 Bi(z) \tag{7.34}$$

where

$$z = \ell^{-2/3}\ (\ell x - E) \tag{7.35}$$

and $c_2,\ c_3$ are constants.

7.2.3 Image states at X gap of Cu(100)

(i) *Energy expression*

For a complete set of wave functions for the system shown in Fig. 7.7, it is necessary to specify also the form of the wave function for $x < 0$, i.e., inside the crystal. As discussed in §5.3, the two-band NFE model gives a reasonable description of the X gap of Cu(100) (Fig. 5.6), so (5.43) is taken as the crystal wave function $\psi_c(x)$ for $x < 0$.

Applying the standard matching conditions to (5.43) and (7.34) at $x = 0$ and to (7.34) and (7.33) at $x = x_{im}$ gives the following expressions for the surface-state energy

$$\frac{\ell^{1/3}[Ai'(z_1)\ +\ K\ Bi'(z_1)]}{Ai(z_1)\ +\ K\ Bi(z_1)} = \frac{(2\alpha)^{-1}W'_{\alpha,1/2}\,(y_1)}{W_{\alpha,1/2}(y_1)} \tag{7.36a}$$

and

$$K = \frac{\ell^{1/3} Ai'(z_0) - \gamma Ai(z_0)}{\gamma Bi(z_0) - \ell^{1/3} Bi'(z_0)} \tag{7.36b}$$

where

$$\begin{aligned} \gamma &= q - \frac{\pi}{a} \tan\delta \\ z_0 \equiv z\,|_{x=0} &= -\ell^{-2/3}\, E && (7.37) \\ z_1 \equiv z\,|_{x=x_{im}} &= \ell^{-2/3}(\ell x_{im} - E) && (7.38) \\ y_1 \equiv y\,|_{x=x_{im}} &= (x_{im} - x_0)/\alpha\ . && (7.39) \end{aligned}$$

As can be seen from the above equations, the only parameter appearing in the energy expressions (7.36) is that of the image plane x_0 (Fig. 7.7).

TABLE 7.1		
n	E (eV)	δ_n
1	10.9075	0.225
2	11.2993	0.206
3	11.3914	0.202
4	11.4258	0.200
5	11.4425	0.199

Because of their oscillating nature, the Whittaker functions $W_{\alpha,1/2}(y)$ give rise to $[\alpha]$[1] zeros. Thus, (7.36) is satisfied for many values of E given by

$$E - V_0 = -\frac{1}{4\alpha^2}, \qquad E < V_0 \tag{7.40}$$

from (7.32). Because $\alpha > 0$, it is convenient to write

$$\alpha = n + \delta_n \tag{7.41}$$

where $n = 1, 2, 3, \ldots$ and δ_n represents the quantum defect. Consequently, the energy levels (7.40) form a *Rydberg series*, as a result of the image-type

[1] $[\alpha]$ denotes the integral part of α.

tail of the potential (Echenique and Pendry 1978). It should be noted that, if the asymptotic part of the surface potential barrier retains its coulombic form up to the point $x = 0$, then $E_n = -n^{-2}$ (the one-dimensional hydrogen atom – Löudon 1959). It is therefore apparent that it is the terminating of the potential at $x = x_{im}$ which causes the values of α to be slightly higher than the positive integers.

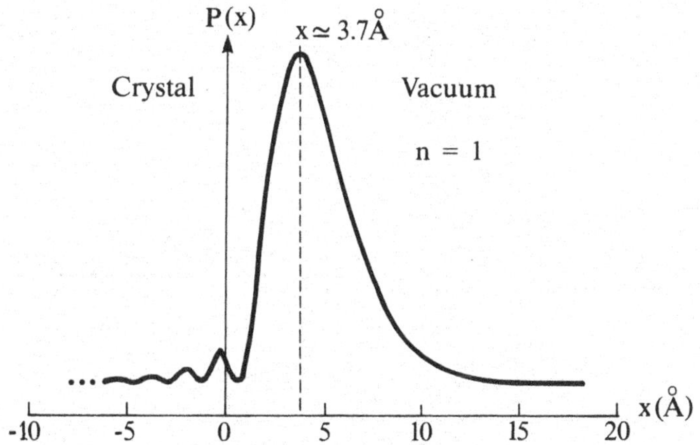

Fig. 7.8. $P(x)$ function for image state $n = 1$, surface being at $x = 0$.

(ii) *Numerical solutions*

The numerical solution of (7.36) leads to a Rydberg series of image states. The energies of the first five levels are given in Table 7.1, where the quantum defect δ_n (7.41) is also shown. The computations were performed for $x_0 = 0.276$Å and $x_{im} = 1.02$Å, which give the best agreement with the recent experimental data ($E_1 = 10.91 \pm 0.02$ eV from the two-photon photoemission measurements of Giesen *et al.* 1987).

(iii) *Localization properties*

The relative probability density functions $P(x) = |\psi(x)|^2/|\psi(0)|^2$ (cf. 3.22) for the first two image states ($n = 1, 2$) are shown in Figs 7.8 and 7.9. As expected, the global maximum of each $P(x)$ is *outside* the crystal. Note that n corresponds to the number of maxima in the vacuum. A comparison of the basic properties of surface states for the step-potential (abrupt-potential termination) and image-potential barriers is given in Fig. 7.10.

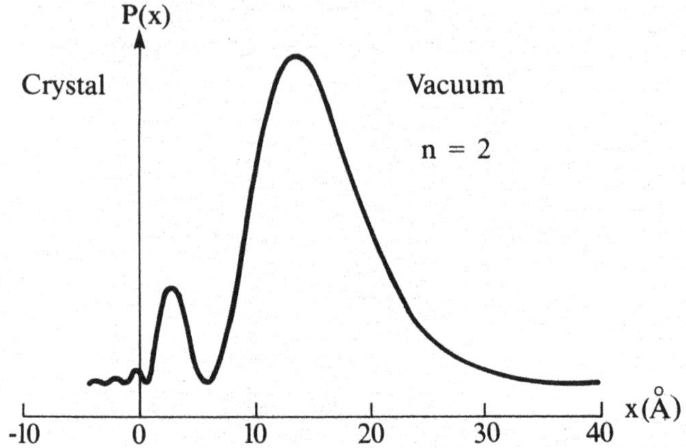

Fig. 7.9. $P(x)$ for image state $n = 2$.

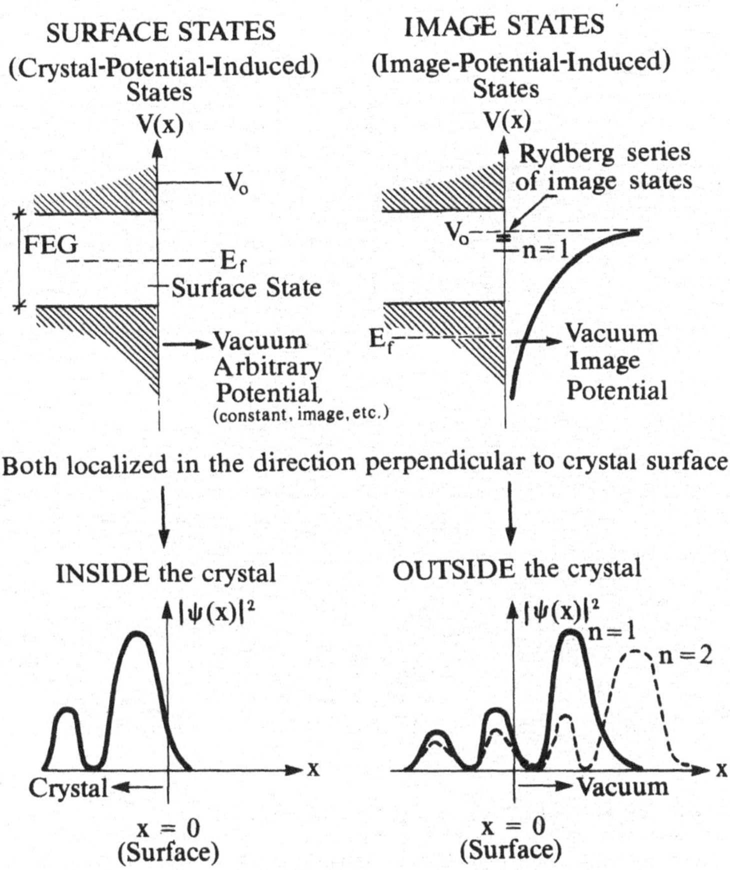

Fig. 7.10. Diagram showing comparison of surface-state properties for step and image potentials.

8

GREEN-FUNCTION METHODS

Learn of the green world what can be thy place
In scaled invention or true artistry.
— Ezra Pound

Of all the methods discussed so far, none is as mathematically sophisticated and elegant as the Green-function (GF) ones to be described here in their various forms. Because these techniques are both powerful and versatile, it is important to master them, despite their more abstract formalism, which tends to make them somewhat more difficult to understand.

The great advantage of GF techniques is that they provide direct access to the density of states (DOS) to which many physical properties are related. For example, the intensity of photoemission from solids is proportional to the DOS. Indeed, most spectroscopic yields are linked to the initial DOS. Other properties, such as bond order (Phillips 1969), can also be related to the GF.

8.1 Resolvent Technique

Although the resolvent method has a long history, dating back to the pioneering work of Lifshitz (1947), Baldock (1952) and Koster and Slater (1954),[1] its present projection operator formulation is due to Koutecký (1965, 1976), who has been the greatest exponent of the technique. The formalism presented here reflects the evolution of the method, and is based on the description given by Davison and Levine (1970).

8.1.1 Formalism

An infinite linear crystal is an unperturbed (non-interacting) system, which may be represented by a cyclic chain of atoms (Fig. 2.8). A finite (or semi-

[1]See also Wannier (1959) for a discussion of the Koster-Slater model.

infinite) crystal can be formed by removing an atom from the chain, thus creating a surface. The removal of the atom perturbs the infinite crystal, so the semi-infinite crystal is a perturbed (interacting) system.

Quantum mechanically, the unperturbed and perturbed systems are described by their effective one-electron Hamiltonians H^0 and H, respectively, which are related by

$$H = H^0 + V \tag{8.1}$$

V being the interaction potential. In the Dirac (1958) ⟨bra|c|ket⟩ notation, the Schrödinger equation for the kth state of the unperturbed system is

$$H^0|\psi_k^0\rangle = E_k^0|\psi_k^0\rangle. \tag{8.2}$$

If the wave function $|\psi_k^0\rangle$ of the infinite crystal is expanded as a LCAO's $|\phi_n^0\rangle$, i.e.,

$$|\psi_k^0\rangle = \sum_n c_{nk}^0|\phi_n^0\rangle \tag{8.3}$$

where $|\phi_n^0\rangle$ describes the motion of an electron in the field of the nth chain atom only, and c_{nk}^0 is the coefficient of the corresponding AO, then inserting (8.3) in (8.2) and multiplying through on the left by $\langle\phi_m^0|$ leads to a system of linear homogeneous equations for c_{nk}^0, viz.,

$$\sum_n (H_{mn}^0 - E_k^0 S_{mn}^0)c_{nk}^0 = 0 \tag{8.4}$$

in which

$$H_{mn}^0 = \langle\phi_m^0|H^0|\phi_n^0\rangle \tag{8.5}$$

and the *overlap matrix*

$$S_{mn}^0 = \langle\phi_m^0|\phi_n^0\rangle = \delta_{mn} \begin{array}{ll} = 1, & m = n \\ = 0, & m \neq n \end{array} \tag{8.6}$$

when the AO's $\{\phi_n^0\}$ are assumed to form an orthonormal set. Hence, (8.6) in (8.4) gives

$$\sum_n H_{mn}^0 c_{nk}^0 = E_k^0 \sum_n \delta_{mn} c_{nk}^0 = c_{mk}^0 E_k^0$$

which, on introducing the dummy index j, say, becomes

$$\sum_n H_{mn}^0 c_{nk}^0 = \sum_j c_{mj}^0 E_j^0 \delta_{jk} \tag{8.7}$$

or, in matrix notation,

$$\mathbf{H}^0\mathbf{c}^0 = \mathbf{c}^0\mathbf{E}^0 \tag{8.8}$$

where $\mathbf{c}^0$ is a matrix whose columns are eigenvectors of $\mathbf{H}^0$, and $\mathbf{E}^0$ is a diagonal matrix whose elements E_k^0 are the corresponding eigenvalues. Assuming that the unperturbed crystal wave function $\{\psi_k^0\}$ also forms an orthonormal set, and using (8.3), yields

$$\langle\psi_j^0|\psi_k^0\rangle = \sum_{m,n} c_{jm}^{0\dagger} c_{nk}^0 \,\langle\phi_m^0|\phi_n^0\rangle = \delta_{jk}$$

which by (8.6) shows that

$$\sum_{m,n} c_{jm}^{0\dagger}\, c_{nk}^0\, \delta_{mn} = \sum_m c_{jm}^{0\dagger}\, c_{mk}^0 = \delta_{jk} \tag{8.9}$$

so

$$\mathbf{c}^{0\dagger}\,\mathbf{c}^0 = \mathbf{1}, \qquad \mathbf{c}^0 = (\mathbf{c}^{0\dagger})^{-1} \tag{8.10}$$

i.e., $\mathbf{c}^0$ is a *unitary* matrix (Margenau and Murphy 1959).

For the perturbed (interacting) system, (8.8) can be written as

$$\mathbf{H}\,\mathbf{c} = (\mathbf{H}^0 + \mathbf{V})\mathbf{c} = E\mathbf{1}\mathbf{c} \tag{8.11}$$

which is the matrix form of a set of equations, each containing an unknown eigenvalue E for each specified perturbed state. After some rearranging, (8.11) yields

$$\mathbf{c} = (E\mathbf{1} - \mathbf{H}^0)^{-1}\,\mathbf{V}\,\mathbf{c}\,. \tag{8.12}$$

However, (8.8) shows that

$$(E\mathbf{1} - \mathbf{H}^0)^{-1}\mathbf{c}^0 = \mathbf{c}^0(E\mathbf{1} - \mathbf{E}^0)^{-1}$$

so

$$(E\mathbf{1} - \mathbf{H}^0)^{-1}\mathbf{c}^0\,\mathbf{c}^{0\dagger} = \mathbf{c}^0(E\mathbf{1} - \mathbf{E}^0)^{-1}\mathbf{c}^{0\dagger}$$

which, by (8.10), gives the *resolvent* (or *Greenian*) *matrix*

$$\mathbf{G} = (E\mathbf{1} - \mathbf{H}^0)^{-1} = \mathbf{c}^0(E\mathbf{1} - \mathbf{E}^0)^{-1}\mathbf{c}^{0\dagger}\,. \tag{8.13}$$

Thus, (8.13) in (8.12) yields

$$\mathbf{c} = \mathbf{G}\mathbf{V}\mathbf{c} \tag{8.14}$$

or

$$(\mathbf{G}\mathbf{V} - \mathbf{1})\mathbf{c} = \mathbf{0}\,. \tag{8.15}$$

For (8.15) to have non-trivial solutions, i.e., $\mathbf{c} \neq \mathbf{0}$ (Margenau and Murphy 1959)

$$\det(\mathbf{GV} - \mathbf{1}) = 0 \tag{8.16}$$

which, together with (8.13), gives the perturbed solutions in terms of the known unperturbed ones.

The above approach, which was originally developed by Lifshitz (1947), Baldock (1952) and Koster and Slater (1954), was generalized by Koutecký (1965), who utilized the projection operator formalism (Löwdin 1962b). Here an association is made between a *projection operator* P and the *space* p into which it projects. Since the operator P of the system is made up of the operators P_i of the subsystems, it follows that

$$P = \sum_i P_i \,. \tag{8.17}$$

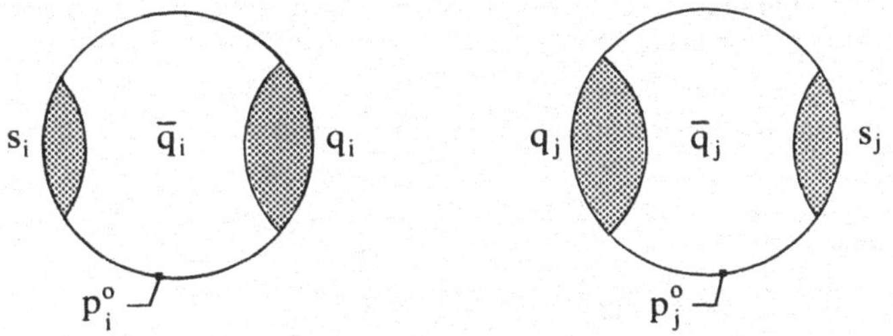

Fig. 8.1. Diagrammatic representation of the *i*th and *j*th subspaces of interacting and non-interacting systems.

The non-interacting and interacting systems can be represented diagrammatically, as shown in Fig. 8.1. For the non-interacting system, the eigenfunctions of the Hamiltonian operator H^0 form the basis of the function space p^0, whose subspaces are p_i^0. In operator terminology, suppose

$$P^0 = \sum_i P_i^0 = I \tag{8.18}$$

I being the *identity* operator (Appendix A). The projection capacity of P^0 can therefore be regarded as *complete*. Moreover,

$$P_i^0 P_j^0 = 0, \quad i \neq j \tag{8.19}$$

i.e., the projection operators P_i^0 and P_j^0 project into the subspaces p_i^0 and p_j^0, which have zero overlap if $i \neq j$.

For the interacting system, the space p of the eigenfunction of H in (8.1) is given by (Fig. 8.1)

$$p = p^0 - s \tag{8.20}$$

so p can be regarded as a subspace of p^0. The removal of the space s from p^0 to form p was exemplified earlier by the removal of an atom from a cyclic chain of atoms, representing an infinite crystal, to form a finite (or semi-infinite) chain of atoms. In general, the space p is the sum of two spaces, namely,

$$p = q + \bar{q} \tag{8.21}$$

where $\bar{q}$ is the space in which *all* the matrix elements of V are zero and q, over which V acts, is the complementary space to $\bar{q}$. Taking the above example further, the formation of a semi-infinite crystal (p) leads to the presence of a surface, which causes the crystal potential to be perturbed in the region corresponding to the space q, leaving the remainder ($\bar{q}$) of the crystal *completely* unperturbed.

In operator language, it is now evident that

$$\bar{Q}V = V\bar{Q} = 0 \tag{8.22}$$

while from (8.18), (8.20) and (8.21) comes

$$I = S + Q + \bar{Q}\,. \tag{8.23}$$

The eigenvalue equation to be solved is (cf. 8.11)

$$H|\psi\rangle = (H^0 + V)|\psi\rangle = E|\psi\rangle \tag{8.24}$$

which can be written as (cf. 8.14)

$$|\psi\rangle = GV|\psi\rangle = GIVI|\psi\rangle \tag{8.25}$$

where the identity operators have been introduced for the purpose of subsequent analysis. Now, since the space s has been removed, it follows that

$$S|\psi\rangle = 0 \tag{8.26}$$

and

$$SH|\psi\rangle = SH^0|\psi\rangle + SV|\psi\rangle = 0$$

which shows that

$$SV = -SH^0 . \qquad (8.27)$$

With the aid of (8.22), (8.26) and (8.27), inserting (8.23) in (8.25) leads to (Koutecký 1965, Koutecký and Davison 1968, Davison and Levine 1970)

$$|\psi\rangle = G(QV - SH^0)Q|\psi\rangle . \qquad (8.28)$$

The retention of the S term in (8.28) enables the initial system to be larger and simpler (e.g., in its symmetry properties) than the system actually being studied. The coefficient equation analogous to (8.14) is (cf. A.2)

$$\langle\phi_m^0|\psi\rangle = \langle\phi_m^0|G(QV - SH^0)Q|\psi\rangle \qquad (8.29)$$

where

$$\begin{aligned}\langle\phi_m^0|\psi\rangle &\neq 0, \quad |\phi_m^0\rangle \in p \\ &= 0, \quad |\phi_m^0\rangle \in s .\end{aligned} \qquad (8.30)$$

The non-trivial solutions of (8.29), via (8.3), require (cf. 8.16)

$$\det\left[\langle\phi_m^0|G(QV - SH^0)Q - 1|\phi_n^0\rangle\right] = 0 \qquad (8.31)$$

the order of the determinantal equation being equal to the number m satisfying the condition $m \in q$. Finally, it should be mentioned that a connection exists (Koutecký and Davison 1968) between the resolvent and partition (Löwdin 1951, 1963, 1964; Levine and Amos 1967) techniques.

8.1.2 Surface states of linear monatomic crystal

Let a finite crystal (p) be created from a cyclic one (p^0) of N atoms ($0 \leq n \leq N-1$) by removing the $n = (N-1)$th atom (s). The projection operator S (Appendix A) associated with the space (s) is given by

$$S = |\phi_{N-1}^0\rangle\langle\phi_{N-1}^0| \qquad (8.32)$$

since s contains only the $(N-1)$th atom, whose AO is $|\phi_{N-1}^0\rangle$.

For a semi-infinite crystal, $N \longrightarrow \infty$, the surface effect at $n = N-2$ is neglected (Grimley 1960), while that at $n = 0$ produces a perturbation in the form of a surface potential V. If this potential perturbs only the first atom at $n = 0$, then the matrix elements of the operator V are given by

$$V_{mn} = \langle\phi_m^0|V|\phi_n^0\rangle = \alpha_0\delta_{m0}\delta_{n0} . \qquad (8.33)$$

Since the space q over which V acts contains only the zeroth atom, the corresponding projection operator is defined by

$$Q = |\phi_0^0\rangle \langle\phi_0^0| \tag{8.34}$$

in the TBA[2].

Inserting (8.32) to (8.34) in (8.29), together with the semi-infinite wave function

$$|\psi\rangle = \sum_{n=0}^{\infty} c_n|\phi_n^0\rangle \tag{8.35}$$

gives

$$\langle\phi_m^0|\sum_n c_n|\phi_n^0\rangle = \langle\phi_m^0|G(|\phi_0^0\rangle \langle\phi_0^0|V$$
$$- |\phi_{N-1}^0\rangle \langle\phi_{N-1}^0|H^0)|\phi_0^0\rangle \langle\phi_0^0|\sum_n c_n|\phi_n^0\rangle$$

which becomes (Feuer 1952, Koutecký 1957)

$$\sum_n c_n\delta_{mn} = \sum_n c_n(G_{m0}V_{00}\delta_{0n} - G_{m,N-1}H_{N-1,0}^0\delta_{0n}) \tag{8.36}$$

where

$$\delta_{mn} = \langle\phi_m^0|\phi_n^0\rangle \tag{8.37}$$

and

$$M_{mn} = \langle\phi_m^0|M|\phi_n^0\rangle \tag{8.38}$$

with $M = G$, V or H^0 for the other matrix elements. On setting the resonance integral

$$H_{N-1,0}^0 = \beta \tag{8.39}$$

and using (8.33), equation (8.36) reduces to

$$c_m = (\alpha_0 G_{m0} - \beta\, G_{m,N-1})c_0\,. \tag{8.40}$$

For $m = 0$ in (8.40), the non-trivial solutions for c_0 require that

$$\alpha_0 G_{00} - \beta\, G_{0,N-1} = 1 \tag{8.41}$$

which may also be obtained from (8.31). Equation (8.41) is the expression for the energy E coresponding to the wave function $|\psi\rangle$ in (8.35).

[2]See Davison and Levine (1970) for a discussion of the next-nearest-neighbour case, and Davison *et al.* (1986) for a treatment of the many-neighbour situation.

For a cyclic crystal, (2.132) and (2.136) in (8.13) give

$$G_{m,N-n} = N^{-1} \sum_k e^{i(m+n)\theta_k}[E - E^0(\theta_k)]^{-1} = G_{m+n} \tag{8.42}$$

since $e^{iN\theta_k} = 1$ by (2.140), which also shows that

$$\delta\theta_k = \theta_{k+1} - \theta_k = 2\pi N^{-1}$$

so, for large N, (8.42) leads to the *Green function*

$$G_m = (2\pi)^{-1} \int_0^{2\pi} e^{im\theta}[E - E^0(\theta)]^{-1} d\theta \; . \tag{8.43}$$

Hence, (8.40) and (8.41) may be written as

$$c_m = (\alpha_0 G_m - \beta\, G_{m+1})c_0 \tag{8.44}$$

and

$$\alpha_0 G_0 - \beta\, G_1 = 1 \tag{8.45}$$

respectively.

The problem has now been reduced to one of evaluating the integral in (8.43), which on setting (cf. 2.131)

$$t = e^{i\theta} \tag{8.46}$$

is transformed into an integral round a unit circle in the complex plane, namely,

$$G_m = (2\pi i)^{-1} \oint t^{m-1}[E - E^0(t)]^{-1} dt \tag{8.47}$$

where

$$E^0(t) = \alpha + \beta(t + t^{-1}) \tag{8.48}$$

from (2.141). After some manipulation, (8.48) in (8.47) leads to

$$G_m = i(2\pi\beta)^{-1} \oint [t^m / F(t)] dt \tag{8.49}$$

where (cf. 2.128)

$$F(t) = t^2 - 2Xt + 1 \tag{8.50}$$

with

$$X = (E - \alpha)/2\beta \; . \tag{8.51}$$

The roots of

$$F(t) = 0 \tag{8.52}$$

are

$$t_{1,2} = X \pm (X^2 - 1)^{1/2} . \tag{8.53}$$

According to the theory of residues (Mathews and Walker 1965), (8.49) can be written as

$$\beta G_m = - \sum_{j=1}^{2} R_j \tag{8.54}$$

where each residue

$$R_j = [(t - t_j)t^m/(t - t_1)(t - t_2)]_{t=t_j} \tag{8.55}$$

is at a pole *inside* the unit circle contour. If any of the t-roots lie *on* the unit circle, then in (8.43) $E = E^0(\theta_k)$ for some *real* value of θ_k. In this case, $t_{1,2} = e^{\pm i\theta_k}$ (8.46) is obtained by putting $X = \cos\,\theta_k$ in (8.53), which by (8.51) gives $E = E^0(\theta_k) = \alpha + 2\beta\cos\theta_k$, as in (2.141) for the cyclic crystal. However, since $E \neq E^0(\theta_k)$ in (8.43), θ must be *complex* (i.e., $\theta = \zeta + i\mu$), which means that $|t_1| = e^{-\mu} < 1$ and $|t_2| = e^{+\mu} > 1$. Hence, the residue R_2 lies *outside* the unit circle contour and must be excluded. Thus, (8.54) and (8.55) yield

$$\beta G_m = - R_1 = t_1^m(t_2 - t_1)^{-1} . \tag{8.56}$$

Inserting (8.56) in (8.45) leads to

$$\alpha_0\beta^{-1} = t_2 = X - (X^2 - 1)^{1/2} \tag{8.57}$$

via (8.53), so

$$X = (\alpha_0^2 + \beta^2)/2\alpha_0\beta \tag{8.58}$$

which by (8.51) gives (Koutecký 1957)

$$E = \alpha + \alpha_0 + \beta^2/\alpha_0 . \tag{8.59}$$

Similarly, (8.56) in (8.44) yields

$$c_m = t_1^m(t_2 - t_1)^{-1}(\alpha_0\beta^{-1} - t_1)c_0 = t_1^m c_0 \tag{8.60}$$

by (8.57). Since

$$t_1 = t_2^{-1} \tag{8.61}$$

from (8.53), equations (8.60) and (8.57) give

$$|c_m/c_0| = |t_2|^{-m} = |\beta/\alpha_0|^m . \tag{8.62}$$

For (8.62) to represent a surface-state wave function, which decays into the crystal as m increases, it is necessary for

$$|\beta/\alpha_0| < 1 . \tag{8.63}$$

Moreover, when this condition holds, comparing (8.59) with (2.142) shows that

$$E > E^0_{\max} \text{ or } E < E^0_{\min} \tag{8.64}$$

so the surface-state energy level (8.59) lies *above (below)* the upper (lower) band edge of the cyclic crystal. It is therefore clear, from the above analysis, that (8.63) is the usual surface-state existence condition (cf. 4.20 and 4.26), which must be satisfied for the wave function and energy level to have the characteristics required of a surface state.

8.2 Dyson Equation Approach

8.2.1 Formalism

Let two semi-infinite crystals be formed from an infinite one by passing a cleavage plane through it in some crystallographic direction. If $H^0(G^0)$ is the Hamiltonian (Greenian) for the unperturbed infinite crystal, and $H(G)$ the Hamiltonian (Greenian) for the perturbed cleaved crystal consisting of the two semi-infinite crystals, then a perturbation expansion for

$$G = (E - H)^{-1} = (E - H^0 - V)^{-1} \tag{8.65}$$

may be obtained in terms of

$$G^0 = (E - H^0)^{-1} \tag{8.66}$$

and the *scattering potential* (or perturbation)

$$V = H - H^0 \tag{8.67}$$

by writing (8.65) as

$$(E - H)G = (E - H^0 - V)G = 1$$

i.e.,

$$[(G^0)^{-1} - V]G = 1$$

via (8.66), or

$$G = G^0 + G^0VG \tag{8.68}$$

which is *Dyson's equation* (Taylor 1970).

Consider now a cleaved one-dimensional infinite crystal, where $|m\rangle$ and $|n\rangle$ denote AO's on opposite sides of the cleavage line (Kalkstein and Soven 1971). Since H is the Hamiltonian of the cleaved crystal, any matrix elements between AO's on different sides of the cleavage line must be set equal to zero. Thus,

$$\langle m|H|n\rangle = \langle m|H^0 + V|n\rangle = 0 \tag{8.69}$$

which, of course, means that the resonance integral coupling the two cleaved crystals is zero. From (8.69), it immediately follows that

$$V(m,n) = -H^0(m,n)\,. \tag{8.70}$$

Moreover, because the cleavage line isolates the two halves of the infinite crystal, its introduction changes the potential field experienced by the electrons. If U is the difference in the potentials in the cleaved and uncleaved crystals, which arises, for example, from a change in the electronic distribution near the surfaces, then

$$V(k,k') = U(k,k') \tag{8.71}$$

where k and k' refer to AO's which are both on the same side of the cleavage line.

For the infinite crystal, the TBA gives

$$H^0(k,k') = \alpha\delta_{kk'} + \beta(\delta_{k,k'+1} + \delta_{k,k'-1})\,, \tag{8.72}$$

while, if the perturbation U is restricted to the surface atoms, then

$$U(k,k') = (\alpha_0 - \alpha)\,\delta_{k,k'}\,. \tag{8.73}$$

If the semi-infinite crystal is formed from the infinite one by cleaving a surface between the -1 and 0 atoms in the chain, then (8.72) and (8.73) imply that, of all the $V(m,n)$ terms, only those corresponding to $(m,n) = (0,0)$, $(-1,-1)$, $(0,-1)$ and $(-1,0)$ are non-zero.

The matrix elements of the Greenian operator in (8.68) are given by the set of algebraic difference equations

$$G(m,n) = G^0(m,n) + \sum_{p,q} G^0(m,p)V(p,q)G(q,n) \tag{8.74}$$

where $G(m,n) = 0$, if m and n refer to atoms on the opposite sides of the cleavage line. For $n = m$, the non-zero $V(p,q)$ terms in (8.74) give

$$G(m,m) = G^0(m,m) + G^0(m,0)V(0,0)G(0,m)$$
$$+ G^0(m,-1)V(-1,-1)G(-1,m) + G^0(m,-1)V(-1,0)G(0,m)$$
$$+ G^0(m,0)V(0,-1)G(-1,m)$$

which, since $G(-1,m) = 0$[3] for $m \geq 0$, reduces to

$$G(m,m) = G^0(m,m) + G(0,m)F^0(m) \tag{8.75}$$

where

$$F^0(m) = G^0(m,0)V(0,0) + G^0(m,-1)V(-1,0)\ . \tag{8.76}$$

From (8.75) comes

$$G(0,m) = G^0(0,m) + G(0,m)F^0(0)$$

so

$$G(0,m) = G^0(0,m)[1 - F^0(0)]^{-1}\ . \tag{8.77}$$

Hence, (8.77) in (8.75) gives (Kalkstein and Soven 1971)

$$G(m,m) = G^0(m,m) + G^0(0,m)F^0(m)[1 - F^0(0)]^{-1} \tag{8.78}$$

for $m \geq 0$.

Taking the trace of $G(E)$ in (8.65) gives

$$Tr\ G(E) = \sum_j \langle\psi_j|G|\psi_j\rangle = \sum_j (E - E_j)^{-1} \tag{8.79}$$

ψ_j being the orthonormal eigenfunctions of H and E_j their corresponding eigenvalues. In evaluating the integral in (8.79), the pole at $E = E_j$ gives

[3]This is equivalent to removing the crystal for $m < 0$.

rise to a singularity in the integration path of the GF, which is avoided by introducing the concept of *complex energy* (Haken 1976), i.e.,

$$E \longrightarrow E - is \tag{8.80}$$

where s is a vanishingly small *positive* number (0^+), which ultimately goes to zero. Putting $\xi_j = E - E_j$, the summand in (8.79) may be written as

$$\frac{1}{\xi_j - is} = \frac{\xi_j}{\xi_j^2 + s^2} + \frac{is}{\xi_j^2 + s^2}$$

so

$$Im\ Tr\ G(E) = \sum_j s(\xi_j^2 + s^2)^{-1} . \tag{8.81}$$

Investigation of the right-hand side of (8.81) reveals that

$$\lim_{s \to 0} \frac{s}{\xi_j^2 + s^2} = 0, \qquad \xi_j \neq 0$$

while

$$\lim_{s \to 0} \int_{-c}^{c} \frac{s\, d\xi_j}{\xi_j^2 + s^2} = \lim_{s \to 0} \left[\tan^{-1}\left(\frac{c}{s}\right) - \tan^{-1}\left(\frac{-c}{s}\right) \right] = \pi . \tag{8.82}$$

Contrasting this with the Dirac δ-function definition

$$\int_{-\infty}^{\infty} \delta(\xi_j) d\xi_j = 1$$

shows that

$$\lim_{s \to 0} \frac{s}{\xi_j^2 + s^2} = \pi\, \delta(\xi_j) \tag{8.83}$$

which in (8.81) leads to the definition

$$\rho(E) = \sum_j \delta(E - E_j) = \pi^{-1} Im\ Tr\ G(E) \tag{8.84}$$

for the *total energy density of states* (Raimes 1972). In the AO representation, (8.84) takes the form

$$\rho(E) = \pi^{-1} Im \sum_m G(m, m) . \tag{8.85}$$

Thus,

$$\rho_m(E) = \pi^{-1} Im\ G(m, m) \tag{8.86}$$

can be interpreted as the *local density of states* (LDOS) at the mth site.

8.2.2 Surface density of states

If the surface atom in a semi-infinite chain is located at $m = 0$, then the *surface density of states* (SDOS) is

$$\rho_0(E) = \pi^{-1}\, Im\, G(0,0) \tag{8.87}$$

via (8.86), with $G(0,0) = G(0)$ given by (8.77), i.e.,

$$G(0) = G^0(0)[1 - F^0(0)]^{-1}\,. \tag{8.88}$$

In order to evaluate $F^0(0)$ in (8.88), it is necessary to know the form of the scattering potential V involved in (8.76), which, in the projection-operator language of second quantization, can be written as

$$V = -\beta(|0\rangle\langle -1| + |-1\rangle\langle 0|) + (\alpha_0 - \alpha)|0\rangle\langle 0| \tag{8.89}$$

where the first term represents the breaking of the β-bond between the $m = 0$ and -1 atoms to form the semi-infinite chain from an infinite one, and the second term denotes the perturbing of the site energy α to α_0 at the surface atom $m = 0$. Thus, in (8.76), the matrix elements

$$V(0,0) = \alpha_0 - \alpha, \qquad V(-1,0) = -\beta \tag{8.90}$$

via (8.89), so

$$F^0(0) = (\alpha_0 - \alpha)G^0(0) - \beta G^0(0,-1)\,. \tag{8.91}$$

Hence, (8.88) becomes

$$G(0) = G^0(0)[1 - (\alpha_0 - \alpha)G^0(0) + \beta G^0(0,-1)]^{-1} \tag{8.92}$$

i.e., the perturbed GF is expressed in terms of the known unperturbed GF and the parameters of the system.

Recalling the GF analysis for the cyclic (infinite) chain, (8.56) and (8.61) show that

$$G^0(n,m) = G^0(|n-m|) = t^{|n-m|}G^0(0) \tag{8.93}$$

where

$$G^0(0) = \beta^{-1}(t^{-1} - t)^{-1}\,. \tag{8.94}$$

Note that, because the atomic chain is infinite and homogeneous, the GF in (8.93) is translationally invariant. With the aid of (8.94), inserting (8.93) in (8.92) leads to

$$G(0) = \beta^{-1}(t^{-1} - 2z)^{-1} \tag{8.95}$$

where (cf. 8.53)

$$t = X + i(1 - X^2)^{1/2}, \qquad |X| < 1 \tag{8.96}$$

$$t = X - \operatorname{sgn} X(X^2 - 1)^{1/2}, \qquad |X| > 1 \tag{8.97}$$

and

$$z = (\alpha_0 - \alpha)/2\beta \tag{8.98}$$

is the dimensionless surface perturbation parameter (cf. 4.8).

As is apparent from the form of (8.79), the reduced energy X_s of the localized surface state is given by the zeros of the denominator of (8.95), i.e.,

$$t^{-1} - 2z = 0 \tag{8.99}$$

which by (8.97) gives (cf. 4.21)

$$X_s = z + (4z)^{-1} . \tag{8.100}$$

Moreover, since $|t| < 1$ in (8.99), it follows that

$$|z| > \tfrac{1}{2} \tag{8.101}$$

or

$$|(\alpha_0 - \alpha)/\beta| > 1 \tag{8.102}$$

via (8.98), which agrees with the previous surface-state existence condition (8.63).

Before turning to the question of finding the SDOS for the perturbed semi-infinite chain, it is useful, for comparison purposes, to first obtain the bulk DOS for the *unperturbed infinite chain.* From (8.93), it is clear that

$$G^0(m, m) = G^0(0) \tag{8.103}$$

so the GF, and the LDOS derived from it via (8.86), is the *same* at *every* atom in the chain. Thus, substituting (8.96) into (8.94) gives

$$G^0(0) = -(2\beta i)^{-1}(1 - X^2)^{-1/2}, \qquad |X| < 1 \tag{8.104}$$

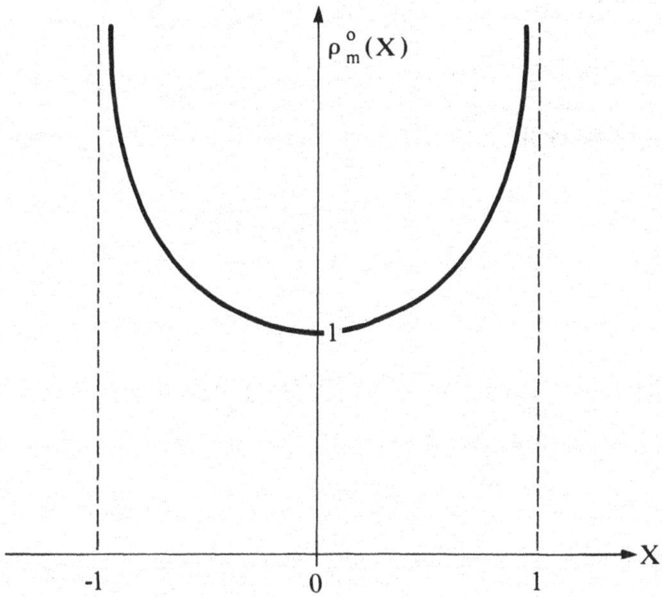

Fig. 8.2. Local density of states at any m site in infinite monatomic chain.

which in (8.86) yields the *site* DOS

$$\rho_m^0(X) = \rho_0^0(X) = (2\pi\beta)^{-1}(1-X^2)^{-1/2}. \tag{8.105}$$

Figure 8.2 displays the schematic curve of $\rho_m^0(X)$ for $2\pi\beta = 1$. The LDOS varies symmetrically from the minimum of unity at the band centre ($X = 0$) to the van Hove singularities (Ziman 1965) at the band edges ($X = \pm 1$). Note that $\rho_m^0(X) = 0$ outside the band, whose width $\Delta X = 2$. In reduced energy notation, it is easy to show that

$$2\beta \int_{-1}^{1} \rho_m^0(X)dX = 1 \tag{8.106}$$

for all m. Thus, the total number of states per atom is unity, as required.

For the *cleaved crystal* situation, the GF at $m = 0$ for $|X| < 1$ can be obtained directly from (8.95) and (8.96), which, together with (8.100), give

$$G(0) = (4\beta z)^{-1}(X_s - X)^{-1}[(X-2z) + i(1-X^2)^{1/2}]. \tag{8.107}$$

Hence, (8.87) and (8.107) yield

$$\rho_0^b(X) = (4\pi\beta z)^{-1}(X_s - X)^{-1}(1 - X^2)^{1/2}, \qquad |X| < 1 \tag{8.108}$$

for the SDOS *inside* the band.

The GF in the surface-state region *above* the band ($X > 1$, $z > 1/2$) is derived by replacing X by $X - is$ (cf. 8.80) in (8.107), which to $O(s)$ leads to

$$G(0) = (4\beta z)^{-1}(A - isB)[(X_s - X) + is]^{-1} \tag{8.109}$$

where

$$A = (X - 2z) - (X^2 - 1)^{1/2} \tag{8.110}$$

and

$$B = 1 - X(X^2 - 1)^{-1/2} . \tag{8.111}$$

From (8.109), it follows that to $O(s)$,

$$Im\ G(0) = -(4\beta z)^{-1}s[A + B(X_s - X)][(X_s - X)^2 + s^2]^{-1} \tag{8.112}$$

which, in the limit as $s \longrightarrow 0^+$, reduces to

$$Im\ G(0) = 0, \qquad X \neq X_s \tag{8.113}$$

and

$$Im\ G(0) = -(4\beta zs)^{-1}A, \qquad X = X_s . \tag{8.114}$$

With the aid of (8.100) and (8.110), inserting (8.114) in (8.87) shows that

$$\rho_0^s(X_s) = (2\pi\beta s)^{-1}[1 - (4z^2)^{-1}] . \tag{8.115}$$

Introducing the function

$$f(X, s) = s[(X - X_s)^2 + s^2]^{-1} \tag{8.116}$$

with the properties

$$\begin{aligned} \lim_{s \to 0^+} f(X, s) &= 0, & X &\neq X_s \\ &= s^{-1}, & X &= X_s \end{aligned} \tag{8.117}$$

of the Dirac δ-function, means that the SDOS (8.115), above the band, can be written as

$$\rho_0^s(X) = (2\pi\beta)^{-1}I_s\ \delta(X - X_s) \tag{8.118}$$

where

$$I_s = 1 - (4z^2)^{-1}, \qquad |z| > \tfrac{1}{2} \tag{8.119}$$

is the *intensity* of the surface state at $X = X_s$. Repeating the same analysis, for $X < -1$ and $z < -1/2$, reproduces (8.118) and (8.119) for the surface state located at $X = -X_s$ *below* the band.

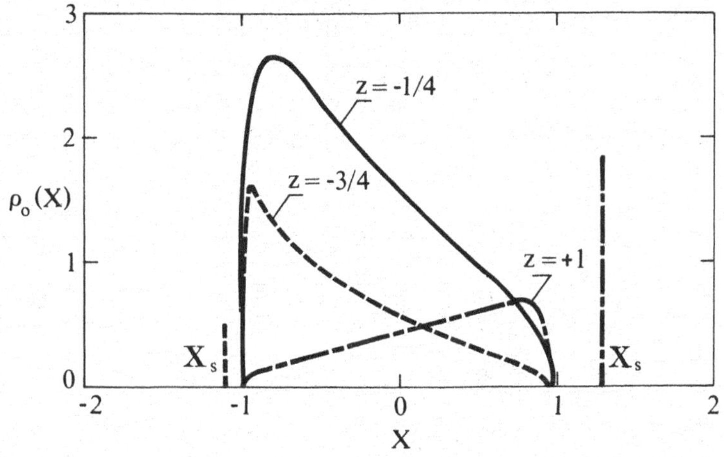

Fig. 8.3. Density of states curves at the surface atom $m = 0$ for values of z indicated. $X = X_s$ locates position of surface state.

Equations (8.108) and (8.118) are the analytical expressions for the band and surface-state contributions to the complete reduced energy spectrum of the SDOS, $\rho_0(X)$, respectively. Graphs of the SDOS curves are drawn in Fig. 8.3 for $2\pi\beta = 1$ and the various values of z indicated. Unlike the uncleaved crystal case (Fig. 8.2), $\rho_0^b(X) = 0$ at the band edges, and varies asymmetrically in the band, its maximum value at $X = X_s^{-1}$ being given by

$$\hat{\rho}_0^b(X_s^{-1}) = (4\pi\beta z)^{-1}[X_s - (2z)^{-1}]^{-1}. \tag{8.120}$$

As z increases from $-3/4$ to 1, the location of the surface state at $X = X_s$ traverses the reduced-energy spectrum from below to above the bulk band. In the case of $z = -1/4$, no surface state exists (8.101), and the area under the bulk-band curve attains its largest value. When $z = 1$, the intensity of the surface-state 'spike' is higher than that for $z = -3/4$. Correspondingly, the area under the bulk-band curve is smaller for $z = 1$ than for $z = -3/4$, in keeping with the sum rule (cf. 8.106)

$$2\beta \int_{-\infty}^{\infty} (\rho_0^b + \rho_0^s) dX = 1 . \tag{8.121}$$

So far, the analysis has been confined to the surface-atom case $m = 0$. However, general expressions for the GF and LDOS at the mth atomic site in the crystal can also be obtained. Equations (8.90) and (8.93) in (8.76) give

$$F^0(m) = t^m G^0(0)(\alpha_0 - \alpha - \beta t) \tag{8.122}$$

which, with (8.93), (8.94) and (8.98), enables (8.78) to be expressed in the form

$$G(m,m) = [1 - t^{2m}(t - 2z)(t^{-1} - 2z)^{-1}]G^0(0). \tag{8.123}$$

Using (8.94) and (8.96), this becomes (cf. Kalkstein and Soven 1971)

$$G(m,m) = i[2\beta(1 - X^2)^{1/2}]^{-1}(1 + t^{2m}T^*T^{-1}) \tag{8.124}$$

where

$$T = (1 - X^2)^{1/2} + i(X - 2z) \tag{8.125}$$

with $|X| < 1$ and $|z| < 1/2$. The mth site bulk LDOS $\rho_m^b(X)$ can now be obtained from (8.124) by means of (8.86).

The GF for the surface-state part of the LDOS is found from (8.123) via (8.94) and (8.97), for which $|X| > 1$ and $|z| > 1/2$. As before, replacing X by $X - is$ and performing some tedious algebra, eventually leads to

$$\rho_m^s(X) = (2\pi\beta)^{-1}(2z)^{-2m} I_s \, \delta(X - X_s) \tag{8.126}$$

which agrees with (8.118) for $m = 0$. Hence, the *total* density of bound states per surface atom is

$$\rho^s(X) = \sum_m \rho_m^s(X) = (2\pi\beta)^{-1} \, \delta(X - X_s) \tag{8.127}$$

by (8.119) and (8.126). Note that (8.121) is now valid for any m and that, since $|2z| > 1$ in (8.126), the surface-state DOS decreases as m increases into the crystal.

8.3 Factorization Method

Garcia-Moliner and Rubio (1969) were the first to propose the so-called *factorization method*, which was later independently extended by Inglesfield (1971) and Velický and Bartoš (1971). From a pedagogical point of view, the original papers describing the method tend to be rather abstruse. Consequently, a simple one-dimensional treatment is furnished here, which follows and develops that of Glasser (1977).

8.3.1 Formalism

Consider a 1-dimensional infinite crystal composed of two semi-infinite ones joined at $x = 0$. Let the crystal potential $V(x)$ be constructed from two different *infinite* periodic potentials $V_+(x)$ and $V_-(x)$, each of which is symmetric about $x = 0$, by writing

$$V(x) = \theta(x)V_+(x) + \theta(-x)V_-(x) \tag{8.128}$$

$$\begin{aligned} \theta(x) &= 0, \quad x < 0 \\ &= 1, \quad x \geq 0 \end{aligned} \tag{8.129}$$

being the *unit step function.* For such a potential, the Schrödinger equation in atomic units $\hbar = 2m = 1$ is

$$[\nabla_x^2 + E - V(x)]\psi(x) = 0 \tag{8.130}$$

whose solutions are subject to the *unmixed* boundary conditions (Appendix B)

$$\psi(\pm\infty) = 0 \tag{8.131}$$

and the requirement that $\psi(x)$ and $\psi'(x)$ are continuous at $x = 0$.

The GF's for the two infinite potentials are defined by the two equations (B.2)

$$[\nabla_x^2 + E - V_\sigma(x)]G_\sigma(x, x') = -\delta(x - x') \tag{8.132}$$

in which $\sigma = +$ or $-$. Strictly speaking, the GF's should, of course, contain the energy E and be written $G_\sigma(x, x'; E)$, but the E is suppressed for brevity. Multiplying (8.130) by $G_\sigma(x, x')$ and (8.132) by $\psi(x)$ and subtracting the resulting equations leads to

$$\begin{aligned} \psi''(x)G_\sigma(x, x') - G''_\sigma(x, x')\psi(x) + [V_\sigma(x) - V(x)]\psi(x)G_\sigma(x, x') \\ = \psi(x)\delta(x - x') \end{aligned} \tag{8.133}$$

which, by (8.128) and (8.129), becomes

$$\psi''(x)G_\sigma(x,x') - G''_\sigma(x,x')\psi(x) + \sigma\theta(-\sigma x)\Delta V(x)\psi(x)G_\sigma(x,x') = \psi(x)\delta(x-x') \qquad (8.134)$$

where

$$\Delta V(x) = V_+(x) - V_-(x). \qquad (8.135)$$

Integrating (8.134) by parts for x from 0 to ∞, gives[4]

$$\begin{aligned} \psi(0)G'_\sigma(0,x') &- G_\sigma(0,x')\psi'(0) + \sigma\int_0^\infty \theta(-\sigma x)\Delta V(x)\psi(x)G_\sigma(x,x')dx \\ &= \int_0^\infty \psi(x)\delta(x-x')dx = \int_{-\infty}^\infty \psi(x)\theta(x)\delta(x-x')dx \\ &= \psi(x')\theta(x') \end{aligned} \qquad (8.136)$$

via (8.129) and the property of the Dirac (1958) δ-function. Similarly, integrating (8.134) from $-\infty$ to 0 yields[4]

$$\begin{aligned} G_\sigma(0,x')\psi'(0) &- \psi(0)G'_\sigma(0,x') + \sigma\int_{-\infty}^0 \theta(-\sigma x)\Delta V(x)\psi(x)G_\sigma(x,x')dx \\ &= \int_{-\infty}^0 \psi(x)\delta(x-x')dx = \int_{-\infty}^\infty \psi(x)\theta(-x)\delta(x-x')dx \\ &= \psi(x')\theta(-x'). \end{aligned} \qquad (8.137)$$

Putting $\sigma = +$ in (8.136) and $\sigma = -$ in (8.137), and combining the equations so obtained, leads to

$$\psi(0)G'_\sigma(0,x') - G_\sigma(0,x')\psi'(0) = \sigma\psi(x')\theta(\sigma x') \qquad (8.138)$$

because by (8.129) the step functions in the left-hand side integrals are zero in the respective ranges of integration.

Taking the limits $x' \to 0^\pm$ in (8.138), where the derivatives $G'_\sigma(0,x')$ are discontinuous, gives the two equations

$$\sigma\psi(0) = \psi(0)G'_\sigma(0,0^\sigma) - \psi'(0)G_\sigma(0,0) \qquad (8.139)$$

[4]Here, and in obtaining (8.137), use is made of the expression $\lim_{x\to\pm\infty} W[G_\sigma(x,x'), \pm\psi_{2,1}(x)] = 0$ with $G_\sigma(x,x')$ defined in (B.15), which is applicable, because (8.131) is an unmixed boundary condition.

$$0 = \psi(0)G'_\sigma(0,0^{-\sigma}) - \psi'(0)G_\sigma(0,0) \qquad (8.140)$$

depending on which combination of $x' = 0^\pm$ and $\sigma = \pm$ is taken. Now, adding (8.139) and (8.140) yields

$$\sigma\psi(0) = \psi(0)[G'_\sigma(0,0^\sigma) + G'_\sigma(0,0^{-\sigma})] - 2\psi'(0)G_\sigma(0,0). \qquad (8.141)$$

However, since the infinite potentials $V_\pm(x)$ are symmetric about $x = 0$, it follows that the corresponding GF's in (8.132) are also symmetric about the discontinuity at $x = x' = 0$, i.e.,

$$G_\sigma(x,x') = G_\sigma(-x,-x')$$

which on differentiating and taking the appropriate limits gives

$$G'_\sigma(0,0^+) + G'_\sigma(0,0^-) = 0. \qquad (8.142)$$

In view of this, (8.141) reduces to

$$\sigma\psi(0) = -2\psi'(0)G_\sigma(0,0) \qquad (8.143)$$

which, for $\sigma = \pm$, may be written as

$$\begin{aligned} \psi(0) &= -2\psi'(0)\mathcal{G}_+(E) \\ \psi(0) &= 2\psi'(0)\mathcal{G}_-(E) \end{aligned} \qquad (8.144)$$

where

$$\mathcal{G}_\sigma(E) = G_\sigma(0,0;E) \qquad (8.145)$$

are the *surface Green functions* (SGF). Dividing the equations in (8.144) yields (Garcia-Moliner and Rubio 1969)

$$\mathcal{G}_+(E) + \mathcal{G}_-(E) = 0. \qquad (8.146)$$

Equation (8.146) is the basic 'recipe' for determining the energies E of the surface states localized at $x = 0$, and in essence replaces the usual condition that $\psi(x)$ and $\psi'(x)$ be continuous at $x = 0$. Thus, the problem has been reduced to the truly remarkable situation of finding the GF's of two *infinite* one-dimensional crystals. The above derivation is easily extended to 3-dimensions (Glasser 1977).

Unfortunately, the preceding analysis is confined to the special case, where the surface lies at a symmetry point of both one-dimensional infinite potentials $V_{\pm}(x)$. However, Inglesfield (1971), and independently Velický and Bartoš (1971), extended the theory by removing this symmetry restriction.

In the more general treatment, the substitution

$$G'_{\sigma}(0, 0^{\pm\sigma}) = D_{\sigma}^{\pm\sigma} \tag{8.147}$$

is made in (8.139) and (8.140). Subtracting these equations then gives the derivative discontinuity equation

$$D_{\sigma}^{\sigma} - D_{\sigma}^{-\sigma} = \sigma 1 \tag{8.148}$$

which is invariant to σ, and may be written as

$$D^{+} - D^{-} = 1$$

thus verifying one of the standard properties of GF's. Eliminating $D_{\sigma}^{-\sigma}$ between (8.141) and (8.148) and collecting terms yields

$$\psi'(0)\mathcal{G}_{\sigma} = \psi(0)(D_{\sigma}^{\sigma} - \sigma 1). \tag{8.149}$$

Dividing the equations in (8.149) for $\sigma = -$ and $\sigma = +$ leads to (Inglesfield 1971)

$$(1 + D_{-}^{-})(1 - D_{+}^{+})^{-1}\mathcal{G}_{+} + \mathcal{G}_{-} = 0. \tag{8.150}$$

As in any result involving the matching of the solutions of a second-order differential equation, the presence of the first-order derivatives is to be expected.

From (8.142) and (8.147), the symmetry condition implies that

$$D_{\sigma}^{+} = -D_{\sigma}^{-} \tag{8.151}$$

which, in conjunction with (8.148), gives

$$D_{+}^{+} = -D_{-}^{-} = 1/2. \tag{8.152}$$

Thus, in the symmetrical case, (8.150) reduces to (8.146), as required.

It should be noted that the Velický-Bartoš (1971) equation, viz.,

$$D_{-}^{+}\mathcal{G}_{-}^{-1} - D_{+}^{-}\mathcal{G}_{+}^{-1} = 0 \tag{8.153}$$

is easily obtained from Inglesfield's (8.150) by means of (8.148) (Bartoš and Máca 1976). Equation (8.153) also reduces to (8.146) via (8.151) and (8.152).

While it is apparent, from the above calculations, that the actual GF for the *complete* interface system is not necessary to obtain the surface-state energy equation, for the subsequent DOS calculations a knowledge of the system GF is required. For the single-interface system, equations (8.132) for G_σ are required, together with the interface system one, viz.,

$$G''(x,x'') + [E - V(x)]G(x,x'') = -\delta(x - x''). \qquad (8.154)$$

If x'' is kept fixed in the negative region, $G(x,x'')$ becomes a function of x only, which may be located in the negative or positive region. When $-\infty < x \leq 0$, equation (8.154) applies solely to the negative region. The only difference between G and G_- is that they satisfy different boundary conditions at $x = 0$.

Multiplying (8.154) by G_- and (8.132), for $\sigma = -$, by G and subtracting leads to an expression containing the potential term $V_+(x)\theta(x)$, which is eliminated by integrating over the negative region to give

$$\begin{aligned} &G(0,x'')G'_-(0,x') - G'(0,x'')G_-(0,x') \\ &= G_-(x'',x')\theta(-x'') - G(x',x'')\theta(-x') \end{aligned} \qquad (8.155)$$

where the limit evaluation at infinity again vanishes due to (8.131). Taking $x' \to 0^-$, and keeping x'' in the negative region, yields

$$G(0,x'')G'_-(0,0^-) - G_-(0,0)G'(0,x'') = G_-(x'',0) - G(0,x'') \qquad (8.156)$$

which via GF symmetry gives

$$G'(0,x'') = G_-^{-1}(0,0)\{G(0,x'')[1 + G'_-(0,0^-)] - G_-(0,x'')\}. \qquad (8.157)$$

Following the same procedure between (8.132), for $\sigma = +$, and (8.154) with integration over the positive region leads to

$$G_+(0,x')G'(0,x'') - G'_+(0,x')G(0,x'') = -G(x',x'')\theta(x') \qquad (8.158)$$

since $\delta(x - x'') = 0$ in (8.154), because x lies in the positive region and x'' is confined to the negative region. On letting $x' \to 0^+$ (i.e., from inside the positive region), (8.158) yields

$$G(0,x'') = -G_+(0,0)G'(0,x'')[1 - G'_+(0,0^+)]^{-1}. \qquad (8.159)$$

After some rearranging, inserting (8.157) in (8.159) shows that the *linkage* GF

$$G(0,x'') = \frac{G_+(0,0)G_-(0,x'')}{\{G_-(0,0)[1-G'_+(0,0^+)] + G_+(0,0)[1+G'_-(0,0^-)]\}}$$

which, on setting $x''=0$ and using (8.145), (8.147) and (8.148), results in

$$\mathcal{G}^{-1} = D_-^+\mathcal{G}_-^{-1} - D_+^-\mathcal{G}_+^{-1}. \tag{8.160}$$

Comparison of (8.153) and (8.160) reveals that the pole of $\mathcal{G}$ gives the surface-state energy, as expected.

In addition to the single-interface system, the interface GF for a two-interface system has been calculated (Bartoš and Velický 1974) to study the electronic structure of a thin film bounded on both sides by a vacuum. The general n-interface problem has also been investigated by Bartoš (1978), using the Dyson equation approach, and by Ueba and Davison (1980), who employed the factorization method. Such calculations are, of course, important in describing the electronic properties of multi-layered systems and superlattices (Bartoš and Davison 1985).

8.3.2 Virtual surface states

For an application of the factorization method, we return to the study of virtual surface states, which were treated previously in §7.1.3. When a linear field is applied to a crystal surface, electrons in surface states are not *truly* localized, since they can 'leak out' of the solid by tunnelling through the potential barrier. For this reason, the term *virtual* surface states was coined for such surface states.

Figure 8.4 displays the potential profile of the model used to study linear field effects on virtual surface states. The crystal occupies the positive half-plane and is represented by the Mathieu sinusoidal potential (§2.5), which, in the limit of small amplitude, approaches the NFE situation (§2.4). The linear field has a positive gradient F and lies in the negative half-plane (Chapter 7). Thus, with the aid of (8.129), the potential for the entire system may be written as (Davison *et al.* 1980)

$$V(x) = (V_0 + Fx)\theta(-x) + 2|V|\cos\, k(x-x_0)\theta(x) \tag{8.161}$$

$$0 \le x_0 \le a \equiv 2\pi/k \tag{8.162}$$

where V_0 is the surface potential, $2|V|$ is the Mathieu potential amplitude, k is the reciprocal lattice constant, a is the lattice constant and the parameter x_0 shifts the crystal potential with respect to the fixed crystal surface at $x = 0$.

The surface-state energy level values can be found by finding the poles of the surface GF $\mathcal{G}$ of the complete system via (8.160), which, for convenience, can be rewritten as

$$\mathcal{G}^{-1} = D_c^+ \mathcal{G}_c^{-1} - D_f^- \mathcal{G}_f^{-1} = 0 \tag{8.163}$$

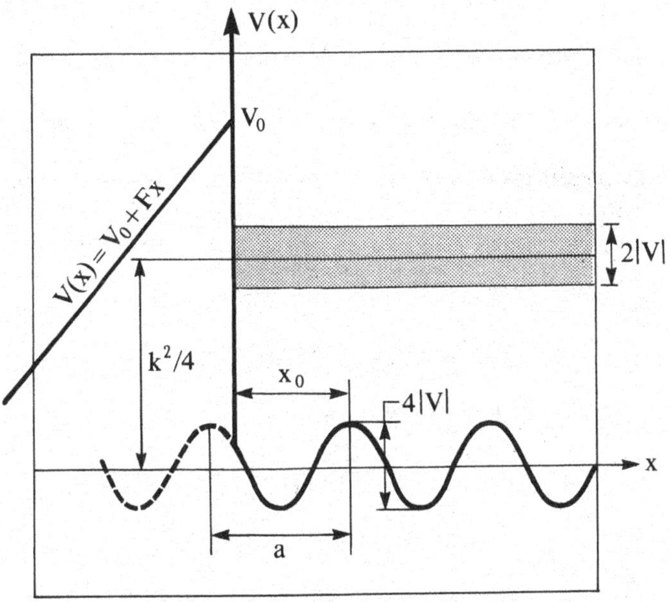

Fig. 8.4. Potential profile of semi-infinite NFE crystal ($x \geq 0$) subject to linear field ($x \leq 0$). The shaded region represents FEG.

where the subscripts c and f designate the crystal and field regions, respectively. The evaluation of each SGF is carried out by means of the direct method described in Appendix B, and leads to the general formula (B.15), viz.,

$$G(x, x') = \frac{\psi_1(x)\psi_2(x')\theta(x' - x) + \psi_2(x)\psi_1(x')\theta(x - x')}{W[\psi_2(x'), \psi_1(x')]} \tag{8.164}$$

where $\psi_{1,2}(x)$ are linearly independent solutions of the Schrödinger equation, and W denotes their Wronskian (Appendix C).

(i) Crystal region

As discussed in §6.2, the Schrödinger equation becomes the Mathieu equation (6.10), for a sinusoidal potential, i.e.,

$$\psi''(z) + (\alpha - 2q \cos 2z)\psi(z) = 0 \tag{8.165}$$

where, for the system depicted in Fig. 8.4,

$$\begin{aligned} z &= k(x - x_0)/2, \quad q = 4|V|/k^2 \\ \alpha &= 4\epsilon/k^2, \quad \epsilon = E - k^2/4. \end{aligned} \tag{8.166}$$

z, α and q are dimensionless quantities, and ϵ measures the energy from the FEG centre. In the mth FEG, the surface-state solutions of (8.165) are

$$\begin{aligned} \psi_1(z) &= \phi(-z)e^{\mu z} \\ \psi_2(z) &= \phi(z)e^{-\mu z} \end{aligned} \tag{8.167}$$

where [cf. (6.8), (6.9) and (6.19)]

$$\phi(\pm z) = \alpha_m ce_m(z, q) \mp \beta_m se_m(z, q) \tag{8.168}$$

$$\mu = 2[(a_m - \epsilon)(\epsilon - b_m)]^{1/2}/mk^2 \tag{8.169}$$

and [cf. (6.24)]

$$\alpha_m = \beta_m[(\epsilon - b_m)/(a_m - \epsilon)]^{1/2}. \tag{8.170}$$

a_m and b_m are eigenvalues corresponding to the cosine and sine elliptic functions $ce_m(z, q)$ and $se_m(z, q)$, respectively. With the aid of (8.166) and (8.167), equation (8.164) for $x > x'$ gives

$$G_c(x, x') = \phi(z)\phi(-z')e^{-\mu(z-z')}\theta(x - x')/\Delta(z') \tag{8.171}$$

$$\Delta(z') = k\{W[\phi(z'), \phi(-z')] + 2\mu\phi(z')\phi(-z')\}/2. \tag{8.172}$$

Differentiating (8.171) with respect to x leads to

$$G_c'(x,x') = [e^{-\mu(z-z')}/\Delta(z')]\{\tfrac{1}{2}k[-\mu\phi(z)\phi(-z')\theta(x-x')$$
$$+\,\phi'(z)\phi(-z')\theta(x-x')] + \phi(z)\phi(-z)\delta(x-x')\} \qquad (8.173)$$

so, as $x \to 0^+$ and $x' = 0$, (8.173) divided by (8.171) yields

$$\frac{G_c'(0^+,0)}{G_c(0^+,0)} = \frac{D_c^+}{\mathcal{G}_c} = \frac{k}{2}[-\mu + \frac{\phi'(z_0)}{\phi(z_0)}] \qquad (8.174)$$

where

$$z_0 = z|_{x=0} = -kx_0/2 \qquad (8.175)$$

from (8.166). Using (8.168), equation (8.174) becomes

$$\frac{D_c^+}{\mathcal{G}_c} = \frac{k}{2}[-\mu + \frac{\alpha_m ce_m'(z_0,q) - \beta_m se_m'(z_0,q)}{\alpha_m ce_m(z_0,q) - \beta_m se_m(z_0,q)}]. \qquad (8.176)$$

In the NFE approximation, $q \to 0$, so for narrow-gap semiconductors (Fig. 8.4)

$$ce_m(z_0,q) \to \cos\, mz_0, \quad se_m(z_0,q) \to \sin\, mz_0 \qquad (8.177)$$

and, for the first FEG $m = 1$, (8.169) and (8.170) can be written as

$$\begin{aligned} \mu &= 2(V^2-\epsilon^2)^{1/2}/k^2 \\ \lambda &= \alpha_1/\beta_1 = [(|V|+\epsilon)/(|V|-\epsilon)]^{1/2} \end{aligned} \qquad (8.178)$$

where $a_1 = |V|$ and $b_1 = -|V|$ (Fig. 2.14a). After some manipulation, equations (8.175), (8.177) and (8.178) in (8.176) give (Velický and Bartoš 1971)

$$\frac{D_c^+}{\mathcal{G}_c} = -\frac{(V^2-\epsilon^2)^{1/2}}{k} + \frac{k}{2}\left(\frac{\epsilon\,\sin\, kx_0 - (V^2-\epsilon^2)^{1/2}\cos\, kx_0}{|V| + \epsilon\,\cos\, kx_0 + (V^2-\epsilon^2)^{1/2}\sin\, kx_0}\right) \qquad (8.179)$$

where $|\epsilon| \le |V|$ for the zeros of (8.163), i.e., surface states in the FEG. In the case of $x_0 = 0$, the sinusoidal potential has a maximum at the surface $x = 0$, and (8.179) reduces to (Garcia-Moliner and Rubio 1971)

$$\left.\frac{D_c^+}{\mathcal{G}_c}\right|_{x_0=0} = -(V^2-\epsilon^2)^{1/2}\left[\frac{1}{k} + \frac{k}{2(\epsilon+|V|)}\right]. \qquad (8.180)$$

(ii) Linear-field region

As in (7.3), putting

$$F^{2/3}z = Fx + V_0 - E \tag{8.181}$$

enables the Schrödinger equation to be written as (7.4), viz.,

$$\psi''(z) - z\psi(z) = 0 \tag{8.182}$$

whose linearly independent solutions are the Airy functions

$$\psi_1(z) = Ai[z\ \exp(-2\pi i/3)], \qquad \psi_2(z) = Ai(z). \tag{8.183}$$

The reason for choosing this particular pair of solutions is because $\psi_1(z)$ may be written as

$$\psi_1(z) = \exp(i\pi/6)[Bi(z) - iAi(z)]/2 \tag{8.184}$$

whose asymptotic form, as $z \to -\infty$, represents an 'outgoing' wave travelling to the left (Duke 1969), which penetrates the linear field barrier. Thus, for $x' > x$, (8.164) gives

$$\begin{aligned} G_f(x, x') &= \frac{Ai[z\ \exp(-2\pi i/3)]Ai(z')\theta(x'-x)}{W[Ai(z'),\ Ai(z'\exp(-2\pi i/3))]} \\ &= \pi[Bi(z) - iAi(z)]\ Ai(z')\theta(x'-x)/F^{1/3}. \end{aligned} \tag{8.185}$$

Differentiating (8.185) with respect to x using (8.181) yields

$$\begin{aligned} G_f'(x, x') &= \pi\{[Bi'(z) - iAi'(z)]Ai(z')\theta(x'-x) \\ &\quad -F^{-1/3}[Bi(z) - iAi(z)]\ Ai(z')\delta(x'-x)\}. \end{aligned} \tag{8.186}$$

Hence, as $x \to 0^-$ and $x' = 0$, (8.186) divided by (8.185) leads to

$$\frac{G_f'(0^-, 0)}{G_f(0^-, 0)} = \frac{D_f^-}{\mathcal{G}_f} = F^{1/3}\left(\frac{\Omega(z_0) + i\pi^{-1}}{Ai^2(z_0) + Bi^2(z_0)}\right) \tag{8.187}$$

where

$$\Omega(z_0) = Ai(z_0)Ai'(z_0) + Bi(z_0)Bi'(z_0) \tag{8.188}$$

with

$$z_0 = z|_{x=0} = (V_0 - E)/F^{2/3} \tag{8.189}$$

by (8.181). As $F \to 0$, $z_0 \to \infty$, and the Airy functions in (8.187) may be replaced by their asymptotic forms, in which case (8.187) reduces to the familiar result (Davison and Tan 1971) for the vacuum situation, namely,

$$D_v^-/\mathcal{G}_v = F^{1/3} z_0^{1/2} = (V_0 - E)^{1/2}. \tag{8.190}$$

Inserting (8.179) and (8.187) in (8.163) gives the inverse SGF expression

$$\mathcal{G}^{-1} = \mathcal{G}_R^{-1} - i\mathcal{G}_I^{-1} \tag{8.191}$$

where

$$\mathcal{G}_R^{-1} = \frac{k}{2}\left(\frac{\epsilon \sin\ kx_0 - (V^2 - \epsilon^2)^{1/2} \cos\ kx_0}{|V| + \epsilon \cos\ kx_0 + (V^2 - \epsilon^2)^{1/2} \sin\ kx_0}\right) - \frac{(V^2 - \epsilon^2)^{1/2}}{k}$$
$$- F^{1/3}\Omega(z_0)[Ai^2(z_0) + Bi^2(z_0)]^{-1} \tag{8.192}$$

and

$$\mathcal{G}_I^{-1} = F^{1/3}\pi^{-1}[Ai^2(z_0) + Bi^2(z_0)]^{-1}. \tag{8.193}$$

Equation (8.191) enables the local density of the virtual surface state to be obtained from (8.86), i.e.,

$$\rho(E) = \pi^{-1} Im\ \mathcal{G} = \pi^{-1}\mathcal{G}_I^{-1}[(\mathcal{G}_R^{-1})^2 + (\mathcal{G}_I^{-1})^2]^{-1} \tag{8.194}$$

where the terms $\mathcal{G}_R^{-1}$ and $\mathcal{G}_I^{-1}$ give the shift and broadening of the virtual surface state, respectively, the latter being inversely proportional to the lifetime τ. The surface-state energies E_s, corresponding to the peaks of the virtual surface states, are given by the zeros of $\mathcal{G}_R^{-1}$ in (8.192). Since $\mathcal{G}_R^{-1}(E_s) = 0$, it follows from (8.194) that

$$\rho(E) = \pi^{-1}/\mathcal{G}_I^{-1}(E_s) \propto \tau. \tag{8.195}$$

Thus, the peak heights in Fig. 8.9 provide a measure of the lifetime of the virtual surface state.

In Fig. 8.5, the graphs of E_s against F are drawn for various values of V_0 with $x_0 = 0.4\pi$. Here, it is important to realise that the location of E_s is determined as a result of the 'competition' between V_0 and F[5], (Fig. 8.7) which is manifest in z_0 (8.189). The low-field region mirrors the findings of Modinos (1974), where E_s decreases with increasing F.

[5] Values of V_0 and F have been chosen to be within accessible experimental ranges for the units used (i.e., 3.9 $eV \leq V_0 \leq 5.7$ eV and $F \leq 1.0$ $V/\text{Å}$).

Figure 8.5 also shows that E_s increases with V_0, as expected. When $F = 0$ and V_0 is large, the electrons are extremely localized in a surface state. When F is 'switched on', the electrons become more delocalized and enter the so-called virtual surface state. Moreover, when V_0 is large (small) the surface state electrons are less (more) sensitive to the application of the linear field.

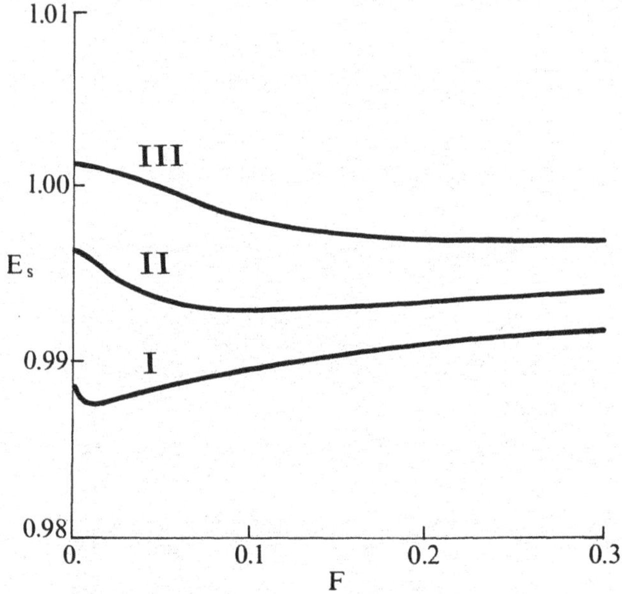

Fig. 8.5. Graph of surface-state energy E_s against field gradient F for I, $V_0 = 1.02$; II, $V_0 = 1.15$; III, $V_0 = 1.3$, when $x_0 = 0.4\pi$ and $a = \pi$.

As F becomes larger, a minimum point is reached on each of the curves I-III, because of the above-mentioned competition between F and V_0. It can be shown that all the minima occur at $z_0 = 0.7$. Actually, if this value of z_0 is inserted in $\mathcal{G}_R^{-1} = 0$ from (8.192), an E_s against F curve is obtained, which passes through all the minima. The minimum in curve III occurs at a larger F value than that of curves of lower V_0 values. Furthermore, as V_0 increases, the minimum continues to move to higher field values and becomes more shallow. After passing through the minimum, E_s increases with F. While the presence of the minimum in the E_s against F curves leads to a smaller shift in the surface-state energy than in previous treatments, it does represent a significant departure from the monotonic variation found there.

For larger values of x_0, all the curves are displaced downwards, while for smaller x_0 values between 0 and $\pi/2$ the curves move upwards until they disappear into the band at a point where their first derivative is zero.

The graphs of E_s against x_0/a are drawn in Fig. 8.6 for various values of V_0 and F with $a = \pi$. The effect of the applied linear field on E_s is demonstrated by considering the two cases $F = 0$ (Velický and Bartoš 1971) and $F = 0.3$. The values of V_0 and F for the curves I-IV are given in Fig. 8.6. Curve V corresponds to an infinite barrier at the surface (Levine 1968), and is also reproduced by plotting the singularities of $\mathcal{G}_R^{-1}$ against x_0/a for any finite value of V_0 and F. The effect of the applied field on the zero-field curve I(IV) is to shift it to the right (left) of the position of curve II(III). The change in the direction of the shift is accounted for by comparing Figs. 8.5 and 8.6. For a given x_0 in Fig. 8.6, the energy

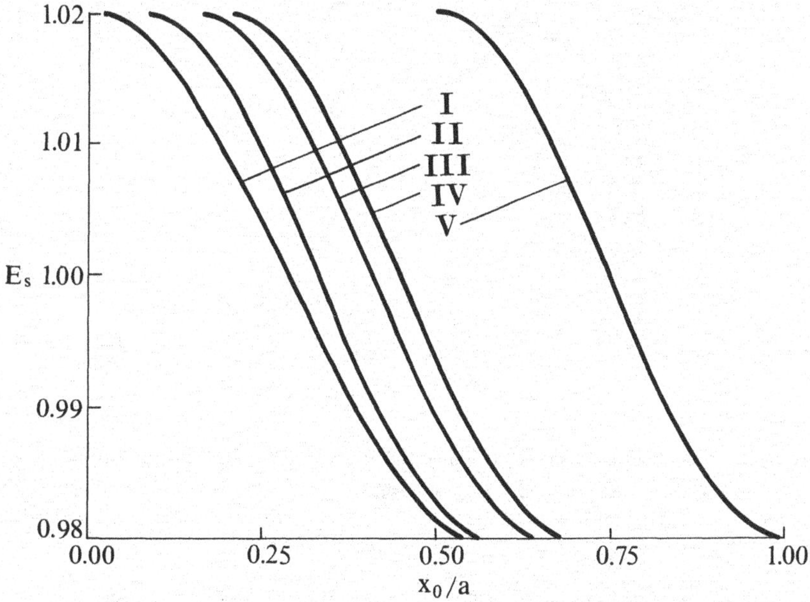

Fig. 8.6. Curves of surface state energy E_s against ratio of surface termination parameter x_0 to lattice constant $a = \pi$. Curve I, $F = 0.0$, $V_0 = 1.02$; curve II, $F = 0.3$, $V_0 = 1.02$; curve III, $F = 0.3$, $V_0 = 1.50$; curve IV, $F = 0.0$, $V_0 = 1.50$. Curve V corresponds to $V_0 = \infty$.

increases (decreases) in going from I(IV) to II(III) as a result of the location of the minimum in Fig. 8.5.

Within the range $0 \le x_0/a < 0.5$, no singularity is present in $\mathcal{G}_R^{-1}(E)$, but, as x_0/a increases from 0.5 to 1, a singularity appears and moves from the conduction band to the valence band along curve V. Any increase in V_0, or a sufficiently large increase in F, shifts E_s towards the singularity (on curve V). Mathematically, the situation at the singularity is one where $V_0 = \infty$ or $F = \infty$. Physically, the presence of the singularity causes E_s to be restricted to a limited region of the energy FEG, the actual limitation depending on the surface termination parameter x_0.

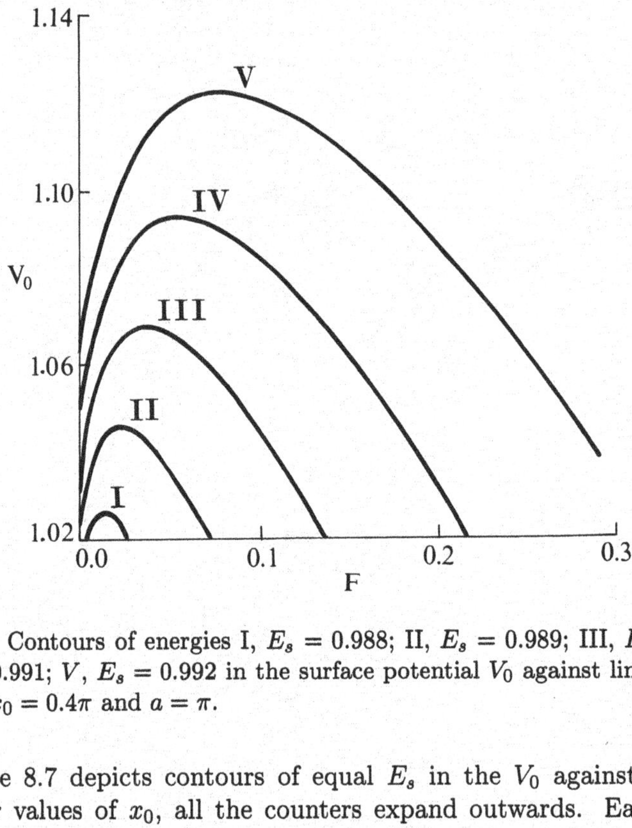

Fig. 8.7. Contours of energies I, $E_s = 0.988$; II, $E_s = 0.989$; III, $E_s = 0.990$; IV, $E_s = 0.991$; V, $E_s = 0.992$ in the surface potential V_0 against linear field F plane for $x_0 = 0.4\pi$ and $a = \pi$.

Figure 8.7 depicts contours of equal E_s in the V_0 against F plane. For larger values of x_0, all the counters expand outwards. Each curve's maximum value of V_0 can be identified with the minimum which was found in the E_s against F curves. The above-mentioned competition between V_0 and F is clearly demonstrated by the 'loop' shape of these curves. Indeed, it is now evident why two values of $F(V_0)$ for one value of $V_0(F)$ gives rise to the same value of E_s.

The plot of $\mathcal{G}_I^{-1}(E_s)$ against F, shown in Fig. 8.8, represents the variation of the broadening of the virtual surface state with the applied field. Although the parameters used are those of curve II, Fig. 8.5, which has a minimum at $F = 0.11$, $\mathcal{G}_I^{-1}(E_s)$ nevertheless increases monotonically with F. The inverse of this curve is proportional to the lifetime τ (8.195) of the electron in this virtual surface state.

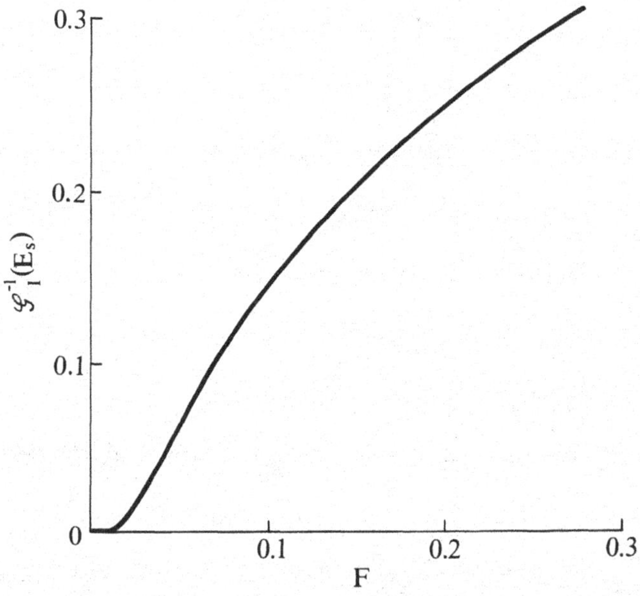

Fig. 8.8. Graph of broadening factor $\mathcal{G}_I^{-1}(E_s)$ as a function of applied field F for $V_0 = 1.15$ and $x_0 = 0.4\pi$.

At first sight, the presence of the minimum in the E_s against F curves, which permits the same E_s value to occur for two different values of F, might seem to be an unexpected result. However, since the effect of increasing the field is to broaden the surface-state energy level (Fig. 8.8), it seems reasonable to believe that the energy DOS will itself be different at two values of F, thus enabling the surface state at these two F-values to be distinguished. Such an explanation is born out in Fig. 8.9, where the DOS curves $\rho(E)$ (8.194) are drawn, again for the parameters of curve II,

Fig. 8.5.

A natural extension of the above work was the study of the virtual surface states of thin films (Jerrard *et al.* 1981), which was based on a knowledge of the two-interface GF (Flores *et al.* 1970, Bartoš and Velický 1974) that led to the general formulation of the n-interface GF (Bartoš 1978, Ueba and Davison 1980).

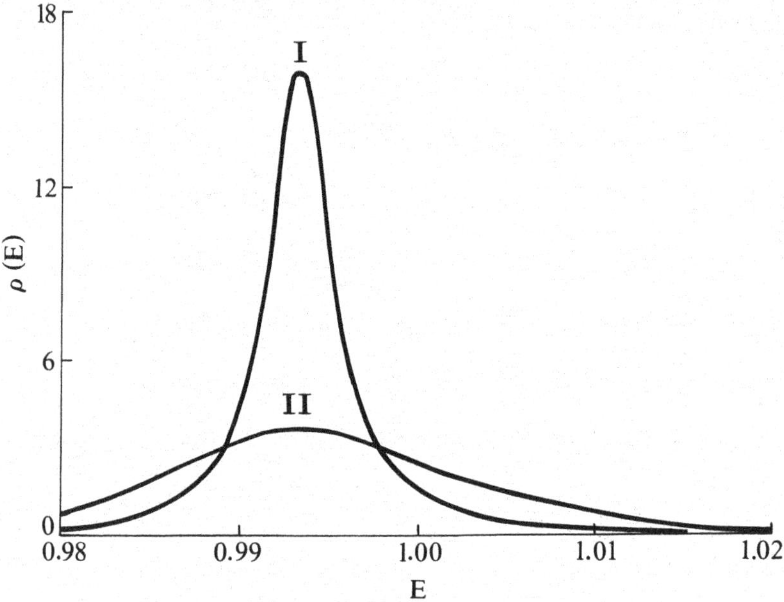

Fig. 8.9. Density of states $\rho(E)$ for virtual surface state when I, $F = 0.05$; and II, $F = 0.227$ with $V_0 = 1.15$ and $x_0 = 0.4\pi$.

9

HISTORICAL DEVELOPMENT

To understand a science, it is necessary to know its history.
— Auguste Comte

After discussing the mathematical aspects of the theory of surface states, it is worthwhile to spend some time tracing their evolution from both a theoretical and experimental point of view. However, a comprehensive review will not be attempted here, but merely an overview, highlighting the various periods of progress. In this way, it is hoped that the reader will gain some historical perspective of the subject. It is interesting to note that, although the theoretical work of Tamm in 1932 predicted the existence of surface states, fifteen years had to elapse before any experimental evidence was uncovered confirming the possibility of their existence.

9.1 Classical Period (1932-49)

Nearly sixty years have now passed since Tamm (1932) first used the KP model to show that certain discrete electronic energy levels of a semi-infinite crystal can exist in that part of the energy spectrum which corresponds to the FEG of the infinite crystal. Tamm also investigated a three-dimensional crystal terminated by a plane, and showed that the density of surface states per cm^2 is $\sim 10^{13}$.

Tamm's ideas immediately drew the attention of other theoretical physicists and, as a consequence, many papers appeared in the 1930's developing and completing Tamm's work. Fowler (1933) discussed the Tamm levels for a finite crystal, and pointed out that the levels should occur in pairs, because of the two faces of the crystal. This was studied in detail by Sokolov (1934), who generalized Tamm's model by introducing finite potential walls and two surfaces. Rijanov (1934) found surface states by considering the eigenfunctions of a thin metal strip. Maue (1935), using Fourier expansions for the lattice field and applying the NFE approximation, was able to indicate the effect of these states on the electrical conductivity of a metal. In 1939, there appeared three very important papers by Goodwin and one by Shockley. Goodwin (1939a), by a suitable extension of Maue's approach, was able to obtain explicit formulae for the wave functions, and the corresponding surface-state energies, in terms of the constants of the crystalline field. Moreover, his methods were applicable to any type of crystal lattice. He first considered a simple cubic crystal, and then extended the results to a general crystal lattice. One of the findings of Maue and Goodwin was that surface states, in any FEG, arise from the allowed bands on either side of the gap. Although this result was obtained from considerations of a finite linear chain of atoms, it may easily be extended to the three-dimensional case.

Both the Maue and Goodwin treatments were based on the assumption that the electrons may be regarded as nearly free. This assumption is not true in practice, of course, the properties of a real crystal being intermediate between those of this extreme case and that of the opposite one represented by the TBA. However, Maue deduced that, according to this latter approximation, surface states could not arise, which would make the visualizing of the intermediate case a matter of considerable difficulty. Consequently, Goodwin (1939b,c) attacked the problem from this opposite standpoint, showed that Maue's deduction was incorrect and introduced a new method of studying surface states, known later as the direct crystal orbital method. Even today, the Goodwin papers provide an excellent basis for those applying either the NFE or the TBA to the study of electronic states of crystal surfaces.

All of the above papers were more methodological in nature than philosophical, and none of them explained how surface states originate from the atomic levels, as the crystal is formed by varying the lattice constant from infinity to a finite value. This was the subject of the classic paper by Shockley (1939), which dealt with a finite one-dimensional potential that was *perfectly periodic throughout the entire crystal, including the outermost unit cells.* The calculations performed by Shockley indicated that the surface levels appeared only when lattice constants were so small that the boundary curves of the allowed energy bands crossed (Fig. 9.3). The levels appeared in the FEG in pairs, one coming from each of the adjoining bands, just as Maue and Goodwin had found earlier in their NFE treatments. Shockley also pointed out that, when the potential in the outermost unit cell is different from that in the interior ones, surface states could occur whether the bands were crossed or not. Consequently, the states obtained by Tamm were a direct consequence of the *asymmetrical* termination of the periodic potential at the surface. In this way, the concept of Shockley and Tamm states was born, and the long controversy about the precise definition of these two types of surface states began.

Shockley also extended his results to three dimensions, where surface levels give rise to surface bands. For diamond, only half of these states will be occupied, so Shockley expected the surface band to be conducting in two dimensions, just as a partially filled three-dimensional band is conducting. Shockley accredited the fact that such a conductivity was not observed to the presence of adsorbed atoms, which lose electrons to the vacant levels in the surface band. This prediction by Shockley gave rise to the discussion – 'Can surface states exist in diamond or not?'

The rather rapid development of the theory of surface states in the 1930's still failed to draw the attention of the experimentalists. This was not surprising, since the natural way of thinking was to believe that surface states, even if they really existed, must produce such subtle effects that the information obtained would not be commensurate with the effort required to secure it. However, the late 1940's were the 'transistor' years, and in many laboratories, particularly those of the Bell Telephone Company, physicists were studying all types of phenomena occurring at crystal surfaces and at metal-metal and metal-semiconductor interfaces. In 1947, a paper was published by Meyerhof on the contact potential difference in silicon crystal rectifiers. The basic idea of the paper was that, in crystal rectifiers, which usually consist of a metal and a semiconductor in contact, the electrons distribute themselves, in such a way, as to align the Fermi levels in both solids. This alignment is always accompanied by a transfer of electrons from the solid with a lower work function to that with a higher one. Now, according to the then current theories, this work function difference was, to a first approximation, equal to the contact potential difference. The aim of Meyerhof's investigation was to measure the contact potential difference and to compare it with the actual work function difference of the separate metal and semiconductor. The results obtained showed *no correlation* between the contact potential and work function differences. These 'anomalous' results of Meyerhof were immediately explained by Bardeen (1947), whose idea was that, if surface states exist and their density is sufficiently high, they will make the work function of a semiconductor almost independent of the height of the Fermi level (E_f) in the interior, which in turn depends on impurity content. If contact is made with a metal, the difference in the work function between the metal and the semiconductor is compensated by the surface-state charge, rather than by the space charge, as was ordinarily assumed. Thus, the space-charge layer is independent of the metal, and there is no reason for the equality between the work function and contact potential differences. The Meyerhof results therefore seemed to be a natural consequence of the possible existence of surface states. But this basic 'existence' problem was still not solved, and Bardeen's fascinating considerations were still only a hypothesis. Two experiments, performed by Shockley – one with Brattain (1947) and the other with Pearson (1948) – finally proved Bardeen's hypothesis and confirmed Tamm's predictions.

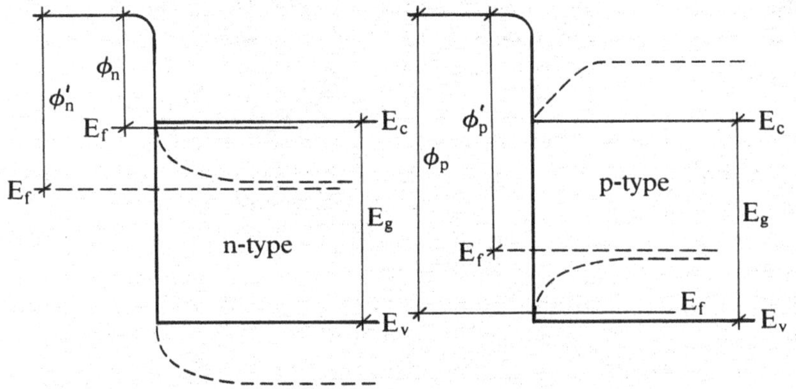

Fig. 9.1. Diagrams showing band bending (broken lines) for *n*- and *p*-type semiconductors. FEG of width E_g is bounded by conduction (E_c) and valence (E_v) band edges (solid lines). Work function in *n*- (*p*-) type sample is $\phi_n(\phi_p)$ relative to Fermi level E_f.

In essence, the experiment by Shockley and Brattain was very simple. Consider two samples of Si, one of *p*-type and the other of *n*-type, chosen in such a way that E_f is located just below the lower edge E_c of the conduction band for the *n*-type and just above the upper edge E_v of the valence band for the *p*-type (Fig. 9.1 - solid lines). In this case, the difference in work functions should almost be equal to the width of the FEG (E_g), i.e., $\phi_p - \phi_n \simeq 1.2$ eV. Of course, the situation will be different, if the existence of surface states is taken into account. Depending on their position with respect to E_f, surface states can act as donors or acceptors, and modify the position of the Fermi level at the surface. In other words, the surface acquires a potential difference with respect to the bulk or, if one prefers, the energy bands bend at the surface (see, for example, Many *et al.* 1965). The direction of bending depends on the position and occupation of the surface states. Consequently, the real situation at the surface is represented by the broken lines in Fig. 9.1. Obviously, $\phi_p' - \phi_n' < 1.2$ eV, so a simple measurement of the work-function difference for such Si samples should show if band bending really occurs near the surface and, thus, if surface states really exist. From their measurements, Shockley and Brattain obtained the value of about 0.6 eV, clearly indicating band bending at the surface and the existence of surface states. It was also possible to estimate from the data that the density of surface states in Si was approximately 10^{14} per cm^2, in good agreement with Tamm's theoretical estimate.

The second experiment, by Shockley and Pearson (1948), demonstrated the existence of surface states even more directly. According to Bardeen's theory, when a charge is induced on the free surface of a semiconductor, some of it goes into the surface states and some into the space-charge region beneath the surface. In the experiment, the charge on the surface was induced by using the semiconductor as one plate of a parallel-plate condenser, and the change in conductance simultaneously determined from a current flow parallel to the plate. Measurements on a number of films of p- and n-type Ge showed that the change in the conductance was correctly predicted by the theory. In other words, it was found that only a fraction of the total induced charge was mobile, indicating that most of it was trapped in surface states. This so-called *field-effect* experiment is still one of the most important tools in the study of surface states.

In the 1940's, very little significant theory was developed, but experimental results showed the important role of surface states in surface phenomena and spurred new interest in the theory of surface states. Consequently, the 1950's can be called the 'theoretical' years, because of the rapid development of the theory of surface states during that period and the complete lack of new experimental information until 1962.

9.2 Tamm, Shockley and Image States

After Shockley published his paper in 1939, the general trend in the literature was to distinguish between Shockley and Tamm surface states. Shockley states were usually defined as the states obtained when a periodic potential was symmetrically terminated at the surface and the atoms were sufficiently close together that band crossing was possible. On the other hand, Tamm states were considered as being entirely due to the change in the potential in the outermost cell of the crystal, whose atoms were far enough apart (i.e., weak interaction), so that no band crossing occurred. Consequently, a convenient method for calculating Tamm states is the TBA, as was first shown by Goodwin (1939a,b). A similar approach was also used by Hoffmann (1951, 1952a,b,c) and Hoffmann and Konya (1951). In the first two papers, an infinite linear chain of similar (Hoffmann and Konya 1951) and different (Hoffman 1951) atoms was studied, and an original way of solving the determinantal (secular) equation was proposed by introducing the so-called *generating function.* This approach can easily be extended to more complex systems, such as AB-type simple square or simple cubic infinite lattices with non-vanishing overlap integrals (Hoffmann 1952b). The secular equation in each of these problems is a determinantal equation, which, for the two- and three-dimensional cases, obviously becomes rather complicated. However, Hoffmann showed that the determinant for the AB cubic lattice can be reduced to the planar case of similar atoms (Hoffmann 1952a), and then readily solved by using Rutherford's (1947) theorem. Finally, Hoffmann (1952c) used the generating-function technique to study the adsorption and surface-state properties of a solid. He considered the crystal as a block, consisting of parallel chains of atoms, each chain being independent of the others. Such a model enables the three-dimensional problem to be reduced to a one-dimensional one, without losing the main features of the three-dimensional treatment. Hoffman's work in one-dimension was extended by Davison and his co-workers to the surface states of general mixed crystals (Davison and Koutecký 1966, Davison *et al.* 1967 and Koutecký and Davison 1968).

Since Shockley states are more likely to occur when the atoms in a crystal are close together and interact strongly, the NFE method should adequately describe this type of surface state. Indeed, as mentioned already, this approach was employed by Maue and Goodwin, even before Shockley's paper appeared. Statz (1950), in his treatment of surface states, discussed two fundamental problems, namely, the mechanism for the formation of Shockley states and the possibility of extrapolating the one-dimensional crystal results to the three-dimensional case. To understand the conditions for the formation of Shockley states in one-dimensional crystals, Statz described the energy spectrum, not as a function of the lattice constant, as Shockley did, but rather in terms of a parameter α, which arises in the theory of Hill's equation. In Statz's approach, therefore, the potential term in the Schrödinger equation is $\alpha V(x)$, where the magnitude of the parameter α characterizes the degree of binding of an electron, and its sign the form

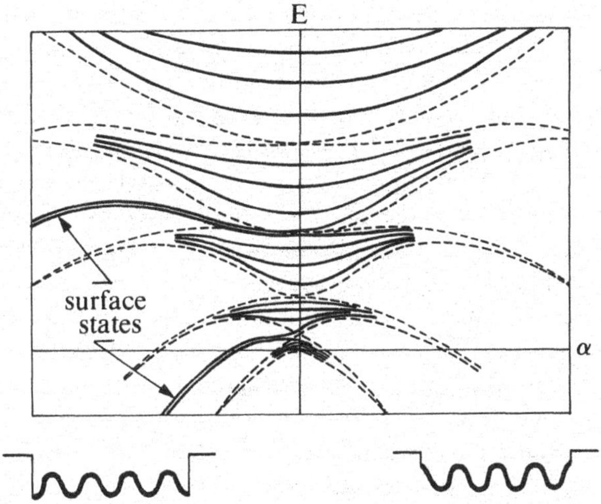

Fig. 9.2. Band structure diagram of E vs α for Mathieu potential, α being electron binding parameter. Broken lines correspond to the band edges. (After Statz 1950.)

of the potential energy. More specifically, Statz adopted the cosine (Mathieu) potential as a special case of the NFE method. The energy spectrum for this potential is shown in Fig. 9.2 as a function of α. It is clear from this figure that there are no surface states in the first FEG for $\alpha > 0$, but they do occur for $\alpha < 0$. The same is true for a square-well potential, which was also considered by Statz. However, says Statz, 'it is only meaningful to cut the crystal between two atoms and, by doing so, one must restrict oneself to the region $\alpha > 0$'. From this it follows immediately that, as long as the lowest energy band is ascribed to the valence band in the one-dimensional case, no surface states exist in the FEG between the valence and conduction bands. Statz also showed that his case included Maue's results as well. The validity of Statz's conclusions can therefore be extended to Goodwin's work, which dealt with a simple cubic crystal, but, in the case in question, the formulation of this problem is quite analogous to Maue's one-dimensional case. Now, as mentioned earlier, the results of the NFE method are unaffected by the transition to the three-dimensional case. Consequently, since no surface states exist in the FEG between the valence and conduction bands, no conducting surface bands can develop. Such a conclusion, however, was an obvious contradiction of Shockley's prediction of a surface conductivity for diamond. This discrepancy was explained by Statz in the following way. Shockley's one-dimensional calculations were performed for an arbitrary potential, using the cellular method (Wigner and Seitz 1933, 1934, Slater 1934, James 1949), and he proposed the following procedure for the extension to the three-dimensional case. If it is possible to write the crystal potential in the form $V(x, y, z) = V(x) + V(y) + V(z)$, then the three-dimensional Schrödinger equation can be separated into three one-dimensional equations, whose solutions are already known. And, on the basis of such an extrapolation, Shockley came to the conclusion that the band of surface states between the valence and conduction bands is only half-filled and, thus, should give rise to a surface conductivity. Statz, however, showed that Shockley's extrapolation is not admissible, because it leads to a completely unreal potential. Furthermore, Statz adopted the cellular method to study the two-dimensional hexagonal lattice with coordination number 3. Such a lattice, as Statz showed, has the same energy spectrum as diamond, so it can justifiably be considered as a two-dimensional analogue to the diamond lattice. Calculations showed that, in the FEG between the valence and conduction bands, surface states exist, but they separate only from the upper empty band, so cannot lead to a surface conductivity. Since these two-dimensional results did not have an analogue in one-dimension, it proved finally that Shockley's extrapolation was not correct. Therefore, there was no reason to expect the existence of conducting surface bands in diamond.

Fortunately, Baldock (1952) was not aware of Statz's paper when he carried out his general investigations of the conditions under which bound states, associated with the atoms in the surface, may exist. These states arise as a result of modifications in the parameters of certain atoms. For the simple cubic lattice, Baldock found the modifications required to produce: (a) surface states associated with all the atoms in the surface, and (b) bound states associated with a particular small group of atoms. He also showed that most of the simple crystal structures do not exhibit surface states without such modifications (i.e., only Tamm states can exist). In some structures, however, Baldock found surface states, even when all the atoms were considered equivalent (i.e., Shockley states). A calculation made for a hypothetical 'monovalent diamond' (including only the nearest-neighbour interaction between p_z - orbitals situated at the lattice sites) yielded the result that Shockley states associated with the (100) and (111) faces existed, one surface state appearing for every two surface atoms, in both cases.

Thus, in 1952 the theoretical situation was very interesting: there was Statz's proof that no conducting surface band can develop in diamond (supported by the fact that surface conductivity had not been observed experimentally in diamond), and Baldock's example showing that a half-filled surface band can still exist. Both considerations were based on highly simplified two-dimensional models, so to explain this discrepancy, a further, more detailed study was needed. In the subsequent years, however, little attention was paid to this problem, and the theory developed in two main directions: further extension of the methods used for surface-state studies and the distinction between Shockley and Tamm states.

Artmann (1952) showed that the TBA, which had been successfully used for studying Tamm states, could also be adopted to treat Shockley states. Obviously, this approximation must now be appropriately modified, because of the strong interaction between neighbouring atoms, which causes the allowed energy bands to broaden and approach each other or even overlap. Consequently, it is not sufficient to treat each energy band as originating from just a single atomic level, though for Tamm states there is no loss of generality in considering the s-states only, as was shown by Goodwin. Artmann, therefore, investigated a crystal model in which hybridization between the atomic orbitals of s and p character was assumed (s- and p-orbitals give rise to the valence and conduction bands in Si, Ge and other semiconductors).

A step towards a sophisticated general theory of localized states was taken by Koutecký (1957), who made use of the Löwdin (1962) projection operator formalism, and generalized the so-called *resolvent technique*, which was originally developed by several other workers (Lifshitz 1947, Baldock 1952, Koster and Slater 1954). This general formalism is based on only two conditions, namely, the applicability of the one-electron approximation, and the assumption that the Hamiltonian of the electron in the crystal, which is confined between two parallel planes, can be written as $H = H_0 + V$, where H_0 is the Hamiltonian of the electron in an infinite ideal crystal and V is the perturbation potential, which is a function only of the distance from one of the planes representing the crystal surface. Although a formal expansion is made in terms of a complete set of Wannier orbitals, the final conclusions of Koutecký's paper are based on the assumption that it is sufficient to use the orbitals of a single band, and that the TBA is adequate for that band.

All the methods mentioned so far have neglected the interaction of the electrons, even within the framework of the Hartree-Fock self-consistent method. However, Tamm (1932), in his classic paper, had pointed out the possible influence of the electron interaction on the occurrence of surface states. An attempt to generalize the method of investigating surface states, in such a way that the condition of self-consistency is satisfied, was made by Koutecký (1958). He proved that the main conclusions of his previous paper (Koutecký 1957), which was based on the one-electron approximation, can be applied to the self-consistent theory of surface states, if, by the concept of a surface state, one means a characteristic function of the Hartree-Fock operator, whose amplitudes tend to zero at distances far from a surface.

The effect of electron correlation on surface states was also studied by Davison and Amos (1965), who employed the method of different orbitals for different spins (DODS) (Löwdin 1962a) to take account of the interaction between electrons inside a linear crystal. Using the TBA, it was found that the electron-electron interaction opened up a so-called *correlation gap* in the middle of the normal one-electron band. The appearance of this FEG made it possible for surface states to occur with energies which had previously been inaccessible to them. Such states, called *inner spin-polarized* surface states, appeared in pairs in the FEG, one being occupied by an α-spin electron and the other by a β-spin electron. Subsequently, Tomášek (1966a, 1966b) extended the work of Davison and Amos by incorporating the DODS method into the resolvent technique to study electron correlation effects in the surface states on the (100) and (111) planes of diamond-like crystals.

Experimentally, the electron paramagnetic resonance measurements of Miller *et al.* (1972) on Si surfaces provided evidence supporting the existence of spin-polarized surface states, though interpretation of the data was, in this case, based on the Hubbard (1963, 1964) electron correlation model.

Although the 1960's saw the rapid development of rather advanced theoretical methods, Aerts (1960) and Phariseau (1960, 1961) returned to the KP potential and used the scattering-matrix and GF methods (Saxon and Hutner 1949), respectively, to study the surface states of one-dimensional semi-infinite crystals. The scattering-matrix technique was also employed by Glasser and Davison (1970) to solve the Dirac equation analytically for the KP potential. This opened the door to the first relativistic treatments of the surface states of heavy atomic solids (Davison and Stęślicka 1969, Stęślicka and Davison 1970, Srinivasan and Davison 1973). In the ensuing discussions, refinements and clarifications of the initial studies were provided by Subramanian and Bhagwat (1972) and Fairbairn *et al.* (1973), the latter being a relativistic GF calculation. Briefly speaking, at sufficient high energies, relativistic effects reveal themselves in the shrinkage of the bulk bands with increasing band number, while the FEG's tend to remain constant in width. Obviously, to accommodate the relativistic effects in this way, the band edges must be relocated. However, since surface states emerge from band edges, it is clear that their non-relativistic energies and existence conditions must be modified in the relativistic picture. Numerous papers have appeared on the various aspects of relativistic surface states (see, for example, Detcheva *et al.* 1990 and references therein). Most recently, Tamura and Feder (1990) adopted a relativistic GF approach in their calculation of the DOS on the Pd(110) surface, and found good agreement with experiment.

Despite the speed at which the methods for studying surface states were evolving, particularly for Shockley states, the definitions of Shockley states (a perfectly periodic potential where the surface states are due to band crossing) and Tamm states (appearing due to the change in the potential in the outermost cell) were still somewhat arbitrary, as was shown by Lippmann (1957). He noticed that the theory of Shockley states becomes identical with that of Tamm states, if it is remembered that the crystal changes its symmetry when the energy bands cross. It is possible to maintain the crystal symmetry, by regarding the boundaries of the unit cell as being shifted instead, leaving an extra half-cell at each face of the crystal. These half cells are 'impurity levels', and are associated with Tamm states. In other words, depending on where the origin of the unit cell is taken, either Shockley or Tamm states can be obtained.

A 'philosophical' discussion, along these lines, was carried on during the next few years, mainly by Koutecký (1960, 1961) and Koutecký and Tomášek (1960a). They studied the properties of Shockley states on the basis of the Artmann model. Since this model was restricted to a very special selection of the Coulomb and resonance integrals, which characterized the problem, Koutecký and Tomášek presented a more general treatment and gave a thorough discussion of the existence conditions, the character of Shockley states and their association with the properties of the initial atomic orbitals. The hybridization of the volume and surface-state wave functions was discussed, and it was suggested that Shockley states were associated with the special bond conditions at the crystal surface, and equivalent to the chemical concept of a free valence at the surface.

Concerning Lippmann's arguments, Koutecký (1960, 1961) pointed out that, as long as the definition of an elementary cell may be arbitrary, Lippmann's consideration was merely formal. Koutecký defined the elementary cell in the model of a linear chain of atoms coupled by alternating strong and weak bonds and, on this basis, discussed the connection between Tamm and Shockley states, and also gave a physical interpretation of the condition for the existence of Shockley states in terms of the unsaturated (dangling) bonds of the surface atoms.

In 1960, there already existed a large variety of methods for studying surface states. Although some of them were quite advanced, they were still rather formal. All applications had been made to such simplified models of a crystal that, even in a rough approximation, they could not define the properties of any real substances. The only two exceptions were the two-dimensional calculations for diamond, performed by Statz (1950) and Baldock (1952), which led to completely divergent results.

Koutecký and Tomášek (1960b) investigated the existence of surface states in diamond and graphite by a simple MO-LCAO method (generalized Artmann model). They showed that in diamond there existed a band of energies pertaining to Shockley states in the FEG between the valence and conduction bands. Since, in the model suggested by Goodwin (1939) and Artmann (1952), the upper and lower bands have one state less than the number of elementary cells, Koutecký and Tomášek assumed that one of the Shockley states originated from the upper and one from the lower band of energies. However, it was still not clear why Statz's calculation led to a different result. The situation was finally clarified by Antončik (1961), who developed a new method for calculating Shockley states, and discussed the general validity of the conclusions obtained on the basis of a one-dimensional crystal, or very simplified models of a real crystal, and its band structure. Antončik (1961) used a simple variation of the orthogonal-plane-wave method, in which the strict orthogonality condition is replaced by a repulsive potential (Antončik 1959a) and the Hamiltonian operator is written in the form $H = -\Delta + V + V_{\text{rep}}$, where Δ is Laplace's operator, V is the potential energy of an electron and V_{rep} is an operator which takes account of the orthogonality condition for the wave function of the valence or conduction band. From the form of the Hamiltonian, it immediately follows that, contrary to Statz's assumption, the potential energy is not determined only by the potential energy V inside the crystal, but rather by the effective potential energy $V + V_{\text{rep}}$. The term V_{rep} can, to some extent, compensate or even exceed the term V (Antončik 1959b); it is not surprising, therefore, that Statz's considerations led to the incorrect conclusion that Shockley states cannot exist in the FEG between the valence and conduction band. Antončik illustrated the general conclusion of his method by the example of a diamond lattice, and proved that a half-filled band of Shockley states exists near the middle of the first FEG. A similar result (a narrow, half-filled band lying somewhat below the middle of the gap) has also been obtained by Pugh (1964), who calculated the surface states of the (111) surface of diamond, based on the linear combination of band orbitals.

Although the papers by Lippmann, Koutecký and Tomášek and Antončik clarified the theoretical picture of surface states in diamond-like structures somewhat, the distinction between the Tamm and Shockley states was still a subject of many discussions.

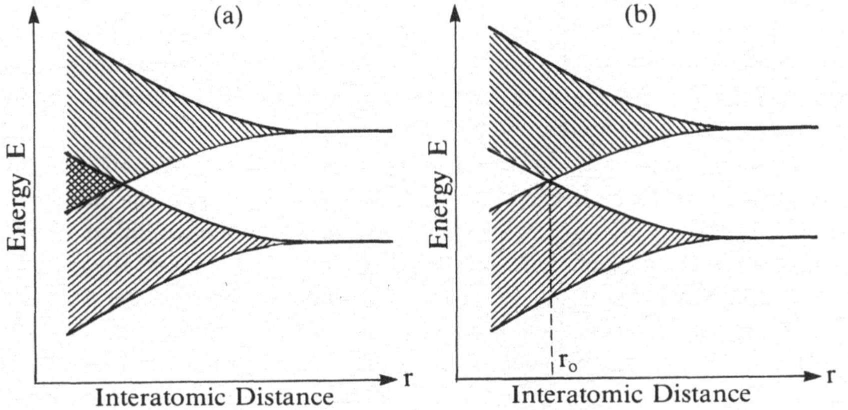

Fig. 9.3. Energy bands of infinite crystal showing, (a) band *overlapping*, (b) band *crossing* with FEG formation when $r < r_0$.

In 1969, Tomášek and Koutecký presented the *explicit* division of surface states on the basis of the following classification of bulk states. In crystals, where the individual energy bands do not interact, the bands are mutually separated, either by FEG's of finite width, or are merely overlapping (Fig. 9.3a). A second group form the crystals in which the electronic states of neighbouring bands interact (Fig. 9.3b) and, as a consequence, the FEG, which is very characteristic of this type of energy spectrum, occurs (this case corresponds to that considered by Shockley). The character of this FEG changes after the intersection. At the point of intersection ($r = r_0$ in Fig. 9.3b), the localized bands appear in the crystal. After the intersection, both bands bear the character of bonding or antibonding states, respectively, in a complete analogy to the hydrogen molecule. In typical intrinsic semiconductors (Si, Ge), the FEG separates the lower (valence) band of predominantly bonding character from the antibonding (conduction) band. Now, for the crystals of the first group (Fig. 9.3a), only Tamm states can exist. For the second group of crystals (Fig. 9.3b), either Tamm or Shockley states can appear, but *never* simultaneously.

A detailed discussion of the behaviour of the Tamm and Shockley states for such 'second group crystals' was given by Foo and Wong (1974),[1] who performed calculations for a semi-infinite, linear chain of atoms with two atomic orbitals (one *s*- and one *p*-like) on each atom, using the GF formalism of Kalkstein and Soven (1971). Although no new conclusions were drawn, they presented a systematic study of the dependence of the energies of both Shockley and Tamm surface states and their existence conditions on parameters, such as the hopping integral and the surface potential perturbation, which was the next step towards a better understanding of these two kinds of surface states.

In subsequent years, these ideas were extended to discuss Tamm and Shockley states in transition metals. For metals, where the number of valence electrons per atom is always less than the coordination number of the atom, it appears that each FEG exists only for a certain region of the surface Brillouin zone. If, in the formation of the FEG, a dominant role is played by two $E(\mathbf{k})$ dispersion relations (corresponding to Bloch functions for different types of orbitals s, p and d), then the FEG can be divided into two groups. When the hybridization interaction is neglected, we have the non-hybridization (NH) gap, and the corresponding dispersion curves should cross (cf. Fig. 9.3a). By switching on Bloch-function mixing, a hybridization (H) gap appears (cf. Fig. 9.3b) and, according to the character of mixing, we can have (s, d), (d, d), (p, d) and (s, p) H-gaps.

Using symmetry arguments, as well as the fundamental properties of chemical bonding in solids, Tomášek and coworkers (Tomášek and Koutecký 1969, Tomášek and Šrubková 1977, Tomášek and Pick 1978) formulated some general rules to predict the existence (or non-existence) of Shockley surface states in any FEG. These rules are: (a) any gap is either an H- or an NH-gap; (b) Shockley states can exist in an H-gap only; (c) in an H-gap of a transition metal, either one or two Shockley states can exist, depending on the number of hybrid orbitals per surface atom cleaved in

[1]Unfortunately, they did not quote the paper by Tomášek and Koutecký (1969).

creating the surface.

In the 1978 Tomášek and Pick article, the starting assumption was 'Shockley states are most important in transition metals'. Consequently, the Shockley model (ideal crystal surface with no surface perturbing potential) was employed to perform a detailed analysis of the existing bulk and surface energy bands of transition metals. Tomášek and Pick obtained a reasonable, qualitative agreement with the results of quantitative surface-state calculations, and thus came to the conclusion that the Shockley model

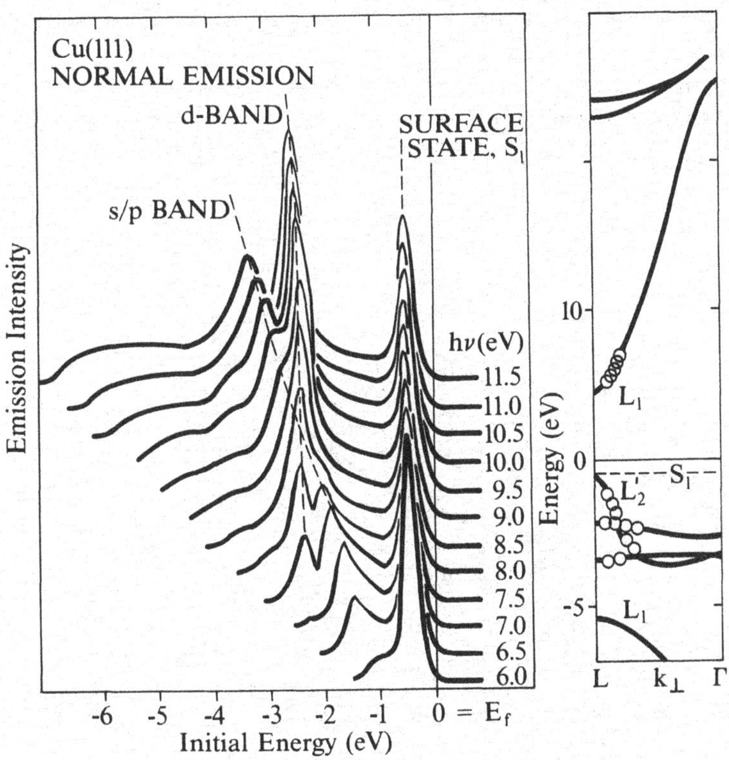

Fig. 9.4. Surface and bulk states for Cu(111). (After Himpsel 1983.)

is basically correct for bcc transition metals. New experimental results extended this conclusion to noble metals, where surface states were observed for Ag(111) and Au(111) (Hansson and Flodström 1978, Hussain and Smith

1978), as well as for the (100), (110) and (112) faces of Au (Heimann *et al.* 1979a) in the *sp*-hybridization gaps.

However, both theoretical and experimental studies of the different faces of Cu soon brought results which could not be explained in terms of the Shockley model. The self-consistent calculations of Gay *et al.* (1979) predicted a large surface-state density near the top of the *d*-band of Cu (100). A few months later, Heimann *et al.* (1979b) observed surface states at or above the top of the bulk *d*-band for Cu(100) and Cu(111). The measured dispersion of the surface-state bands falls in the absolute energy gaps of the projected bulk-band structure for both faces. Since the parent energy gaps were not caused by the hybridization of crossed energy bands, the Shockley model could not be used. Instead, the observed states were interpreted in terms of Tamm surface states. Similar *d*-surface states were also observed on Au(100)-(5×20) and Au(111) surfaces (Heimann *et al.* 1979b) and on Ni(100) (Plummer and Eberhardt 1979).

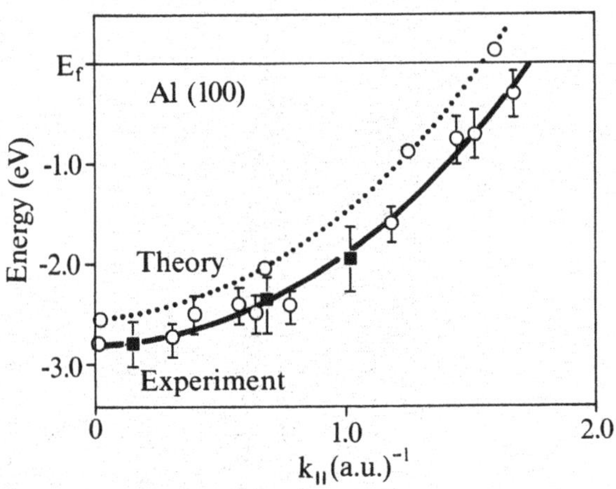

Fig. 9.5. $E(k_{\|})$ dispersion of surface state on A1 (100). (After Himpsel 1983.)

A series of new and interesting results were obtained from angle-resolved photoemission measurements of surface states in metals (Himpsel 1983). In photoemission experiments, the decaying surface states can be distinguished from the oscillatory bulk states, because they lack a well-defined momentum $k_\perp$ perpendicular to the surface. Generally, bulk states exhibit the energy band dispersion $E = f(k_\perp)$, when the photon energy $h\nu$ is varied, while surface states remain stationary in their binding energy, and change only in intensity. Thus, for normal emission, peaks due to surface states show no dispersion in their initial energy as a function of photon energy (Knapp *et al.* 1979), as is illustrated in Fig. 9.4, where surface and bulk states are shown for Cu(111). As can be seen in Fig. 9.5 for Al(100), surface states exhibit a two-dimensional band dispersion $E = f(k_{\|})$.

Both measurements and calculations, performed for different surfaces of transition and noble metals, indicate the existence of *sp*- and *d*-like surface-state bands, which are often referred to as Shockley- and Tamm-type states, respectively. Consequently, in the early 1980's the following 'classification' of surface states in metals was generally accepted (Himpsel 1983). Shockley states are surface states, formed by nearly-free *s* and *p* electrons, which often extend several layers into the solid. On the other hand, *d*-like surface states in transition metals, with atomic-like wave functions localized essentially on surface atoms, are usually considered to be Tamm states.

The use of inverse photoemission for studying *unoccupied* (empty) metal surface states provided new insight into their properties. In 1984, two papers appeared, simultaneously, identifying empty surface states on Cu(100) (Dose *et al.* 1984) and Au(100) and Cu(100) (Straub and Himpsel 1984) as *image* (or image-potential-induced) *states* (see Chapter 7). These states form a special class of unoccupied surface states, whose wave functions ($|\psi|^2$) have maxima *outside* the crystal surface. Although the possible existence of such states was conceived by Shockley in his classic 1939 paper, their physical origin (i.e., of an electron bound to its own image) differs qualitatively from that of Shockley states.

Therefore, the 'real' history of image states began in 1969, when Cole and Cohen, and independently Shikin (1970, 1971), formulated the description of image-potential-induced states for liquid He. They were the first to notice that liquid He totally reflects electrons near the vacuum zero in energy and, thus, should trap electrons in a Rydberg series. A simple one-dimensional approach to this problem for crystals was given by Kolář and Bartoš (1973) and Garcia and Solana (1976). An analysis of inelastic-scattering effects on the Rydberg spectrum was made by Echenique (1976). Rundgren and Malström (1977) used the image potential to construct the surface-barrier model based on the existing theoretical information (MacColl 1939, Cutler and Davis 1964, Jennings 1971, Jennings and Read 1974, McRae and Caldwell 1976, Jennings *et al.* 1988). This model was then employed to perform the low-energy electron diffraction (LEED) calculations, which were soon confirmed by the experimental work of Adnot and Carrette (1977) and McRae *et al.* (1977). A more general discussion of image states, including lifetime broadening, was given by Echenique and Pendry (1978), who also proposed a one-dimensional multiple-reflection model, which was later extended by Smith (1985) and Echenique and Pendry (1989).

Recently, a new experimental technique for studying image states was developed, namely, the high-resolution two-photon photoemission spectroscopy (Giesen *et al.* 1985, Steinmann 1989). The binding energy (E_b) of the image states on Cu(100) and Ag(100) was measured (Giesen *et al.* 1987) and compared with that for the (111) surfaces of Cu, Ag and Ni (Giesen *et al.* 1986). The comparison showed that E_b does not depend on the material of the sample, but changes considerably with the crystal orientation. Consequently, the image states, at virtually every metal surface, seem to have a universal binding energy of $E_b \sim 0.7$ eV below the vacuum level (for the lowest, $n = 1$, image state), being close to 1/16 Rydberg = 0.8 eV, which is expected for a purely Coulombic potential (Giesen *et al.* 1986, 1987). This conclusion, which is also supported by phase-analysis considerations (Giesen *et al.* 1987), rules out the assignment of the observed image state for Ag(100) to the $n = 2$ state, as a result of the surface corrugation (Garcia *et al.* 1985).

At present, both experimental and theoretical studies of image states do not really use the terminology of the Tamm and Shockley states. Instead, they tend to classify these states into two categories (Hulbert *et al.* 1986): (a) *crystal-induced states*, which originate from the crystal termination and are localized *inside* the crystal at the surface and (b) *image-induced states*, which are localized *outside* the crystal, and whose origin is the long-range image potential.

Surface states in semiconductors are mostly interpreted in terms of bond orbitals. On clean semiconductor surfaces, the surface bands often form a band of unsaturated 'dangling bond' surface states somewhere in the FEG. These unsaturated states can lead to instabilities and, as a consequence, the surface atoms tend to rearrange themselves to minimize the energy associated with broken bond states (surface reconstruction). Using LEED technique, Schlier and Farnsworth (1959) were the first to observe that Si(001) was reconstructed. In subsequent years, more and more attention was paid to such surface reconstructed crystals. It is obvious that, when a crystal surface is non-ideal, the distinction between Tamm and Shockley states lacks any physical sense. It is not surprising therefore that, when semiconductor surfaces are investigated, the experimentalists usually use the term 'surface states', though the Tamm and Shockley states terminology is still used in some model calculations for semiconductors with ideal surfaces.

In 1984, there appeared a paper by Zak, which was a continuation of the almost forgotten 'classic' discussion on the distinction between Tamm and Shockley states in a crystal. Zak's considerations were, in principle, similar to those of Lippmann (1957). However, his more general approach and detailed analysis of one-dimensional models led to new and very interesting conclusions. According to Zak, the possible appearance of surface states in a FEG is closely connected to the symmetry of energy bands in the perfect (unperturbed) crystal. The derived symmetry criterion (a selection rule for surface states to appear in the FEG) associates Shockley states with surface states that appear in a crystal whose surface is at a symmetry centre (or a symmetry plane in three-dimensional crystals). When the crystal surface is located at a general (arbitrary) point in the lattice, we have Tamm states. In accordance with Lippmann's results, Zak also showed that band crossing changes the symmetry of the band, so that a FEG, which could not possess a surface state, becomes one that can. Finally, Zak came to a conclusion which, at first sight, seems to be a natural consequence of these considerations, but, in fact, is very new and rather striking. He concludes that, 'since only the symmetry of the band is important, there is, in general, *no need for band crossing for the appearance of Shockley surface states.*' As a beautiful example supporting this statement, Zak pointed out the Mathieu problem (see Chapter 6), where no band crossing is at all possible, but surface states are still regarded as Tamm or Shockley states depending on the termination of the crystal.

So, the story of Tamm and Shockley states is told and, through Zak's clear insight, the controversy of their classification lingers on only in the traditional sense. Thus, since 'bad habits die hard', one can expect to find a residue of the classification in the present work and future literature.

9.3 Energy Distribution of Surface States in Semiconductors

9.3.1 Photoemission measurements on cleaved silicon

It was necessary to wait until the nineteen sixties, for new experimental information to emerge about surface states and their energy distribution in semiconductors. The main difficulty lay in finding a method to obtain atomically clean and atomically smooth semiconductor surfaces. Allen and Gobeli (1962) used, for the first time in surface-state studies, the technique of cleaving monocrystals of Si in ultra-high vacuum ($< 10^{-10}$ Torr.). The principle behind their experiment was that, in the presence of surface states, the position of the Fermi level at the surface (surface potential) differs from that in the bulk (bulk potential). An illustration is given in Fig. 9.6, where the solid lines represent the predicted semiconductor surface behaviour for the case of no surface states, and the broken lines describe the model when surface states are present in the FEG. Obviously, the surface potential is determined by the relation

$$(E_f - E_I)_S = (E_v - E_I)_B - (\Phi - \phi) \tag{9.1}$$

E_f being the Fermi level, E_I the intrinsic level, E_v the top of the valence band, Φ the photoelectric threshold of the semiconductor (the lowest photon energy able to excite electrons from the filled levels over the surface barrier) and ϕ the work function. The difference between the bulk and surface potentials determines the bending of the bands, and hence the charge Q_{sc} in the space-charge layer. Since this is balanced at a free surface by an equal and opposite surface-state charge $Q_{sc} = -Q_{ss}$, the latter is now known over the range of surface potential values, which can be explored by

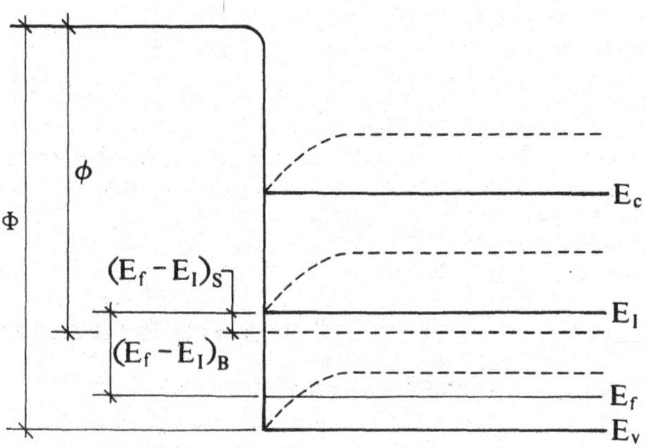

Fig. 9.6. Band structure behaviour in vicinity of semiconductor surface without (solid lines) and with (broken lines) surface states.

varying the bulk doping. Assuming that surface states are not seriously altered by bulk doping, considerable information can be obtained about their density and distribution, since these must satisfy the Q_{ss} vs $(E_f - E_I)_S$ relation mentioned above.

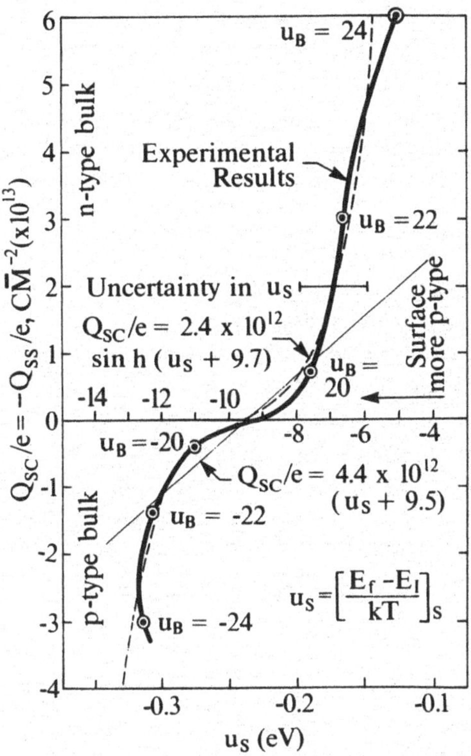

Fig. 9.7. Surface-state charge Q_{ss}/e vs surface potential u_S for Si(111). (After Allen and Gobeli 1962). (After Allen and Gobeli 1962.)

In the experiment, only the (111) surface of Si, i.e., the cleavage plane, was studied. The bulk potential $u_B = (E_f - E_I)_B$ was determined for each surface from its resistivity. The surface potential $u_S = (E_f - E_I)_S$ was found, via (9.1), from the difference between work function ϕ and photoelectric threshold Φ. The value of ϕ was determined by using the Kelvin contact-potential difference method and that of Φ from photoelectric threshold measurements. With the surface potential, bulk potential and impurity concentration known, the charge in the space-charge region Q_{sc} was then computed from the relation (Gobeli and Allen 1962)

$$Q_{sc} = -\frac{\kappa}{4\pi}\left(\frac{du}{dx}\right)_{x=0} \tag{9.2}$$

where $u = (E_f - E_I)/kT$ and κ is the dielectric constant. The results are shown in Fig. 9.7, where the surface-state charge $Q_{ss}/e(= -Q_{sc}/e)$ is plotted vertically and surface potential u_S (in kT units) horizontally. Allen and Gobeli discussed the fit to the data given by the assumed surface charge-surface potential relations, and showed that the best fit over the entire data range is given by the equation (broken curve in Fig. 9.7)

$$Q_{ss}/e = -2.44 \times 10^{12} \sin h(u_S + 9.7). \tag{9.3}$$

The hyperbolic sine relation between Q_{ss} and u_S can be provided by at least three different surface-state distributions: (a) two bands of surface states, one above and the other below the neutral E_0, corresponding to the flat-band condition of the surface, where $Q_{ss} = 0$ (Fig. 9.8a); (b) two discrete levels, one above and one below E_0 (Fig. 9.8b); (c) a hyperbolic cosine distribution of states centered at E_0 (Fig. 9.8c). In deciding which of the above three distributions is most likely, Allen and Gobeli believed that 'either the two-band or the two discrete-level model could apply, depending on the strength of interaction between neighbouring surface atoms. There seems to be no reason to expect the hyperbolic cosine distribution, unless

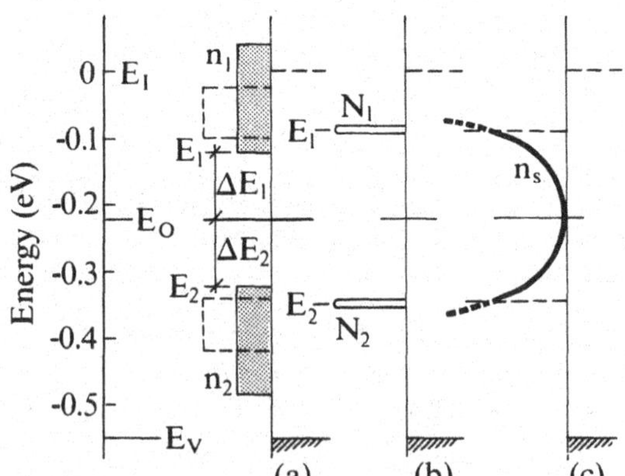

Fig. 9.8. Three surface-state distributions, which give rise to hyperbolic sine relation between Q_{ss}/e and u_S in Fig. 9.7. (After Allen and Gobeli 1962.)

the surfaces are inhomogeneous, and this happens to approximate the result of many different levels.'

The main findings of Allen and Gobeli indicated a surface-state density approximately equal to the surface atom density of $8 \times 10^{14}/\text{cm}^2$ with the states located in two groups: one normally empty just below the center of the FEG and one normally filled just above the valence-band edge. The photoelectric and work function measurements also indicated that the neutral point of the surface states, E_0, lies in the lower half of the FEG. The interpretation of their data required that the cross-section for the interaction of photons with electrons in surface states be so small that photoelectrons from surface states cannot be observed, and that all photoelectrons arise from volume states. Their yield curves were also in good agreement with theoretical predictions (Brust 1964), which supported the assumption that all of the photoelectrons arose from volume states. Nonetheless, Scheer and Van Laar (1963) and Van Laar and Scheer (1965), who obtained data almost identical to those of Allen and Gobeli, assumed that the photoelectrons very near threshold arose from surface states. This interpretation required that the neutral point of the surface states lie slightly above the center of the FEG. This difference of interpretation was resolved by Aspnes and Handler (1966), who obtained the neutral point of surface states by measuring the conductivity of the space-charge region adjacent to the surface. These measurements were performed by a method proposed by Handler (1963), the basic idea of which is shown in Fig. 9.9. As can be seen, the conductance of the arrangement is proportional to the area of the

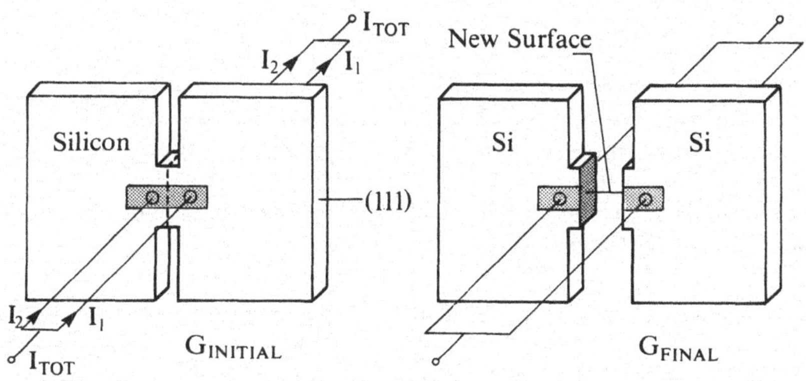

Fig. 9.9. Schematic diagram showing sample and contact used in measuring conductivity of a cleaved surface. (After Handler 1963.)

evaporated contact. After cleavage, the area for current flow has not changed, so that, if no surface conductivity is present, the total conductance will remain unchanged. However, if the cleavage does produce a change then that change must be equal to the surface conductance, $G_s = G_{\text{final}} - G_{\text{initial}}$. G_s may be positive or negative, depending on the position of E_f with respect to the energy levels of the induced surface states, as is illustrated in Fig. 9.10. Prior to cleavage, the region of the crystal around the cleavage plane will certainly be in the flat-band condition, because it is part of the bulk. Immediately following cleavage, but before the surface and volume have come into equilibrium, the energy-level diagram of a *p*-type semiconductor is as represented in Fig. 9.10a, while Fig. 9.10b shows the situation after equilibrium (redistribution of charge between the surface and volume). Since the sample was originally *p*-type, a decrease in conductivity was observed. However, if E_f was just equal to or above the neutral point of the surface states, no change or increase in conductance was observed, respectively (Figs. 9.10c and 9.10d).

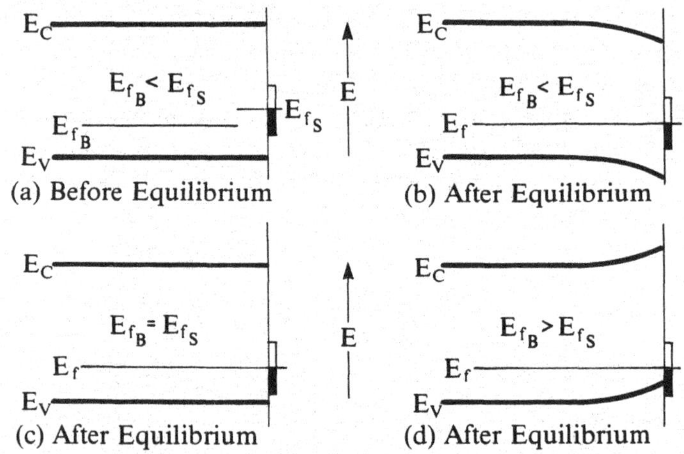

Fig. 9.10. Band bending at semiconductor surfaces before and after cleavage. (After Aspnes and Handler 1966.)

In the experiment, Au doped Si was used, because it has the bulk E_f located slightly below the midgap at room temperature. Thus, a simple measurement of the sign of the change of the conductance can determine whether the center of the surface states is below or above midgap. Aspnes and Handler observed an increase in the conductance, which meant that an accumulation region of holes was formed (in the initially *p*-type crystal) as a result of cleavage. Therefore, the neutral point must lie closer to the valence-band edge than the bulk E_f does (Fig. 9.10d), i.e., below the midgap at 0.28 ± 0.08 eV above the valence-band edge, in very good agreement with the results of Allen and Gobeli. Consequently, the photoelectrons very near to the threshold were due to emission from the valence band, as postulated by Allen and Gobeli, and not from a surface-state band, as postulated by Van Laar and Scheer.

9.3.2 Calculation of surface-state bands in semiconductors with ideal surfaces

One of the most interesting properties of surface states is their energy distribution. However, very little was known about it prior to the measurements of Allen and Gobeli (1962). From theoretical considerations, based on very simplified models, it followed that only *one band* of surface states can be formed in the FEG. Allen and Gobeli's experiment suggested a *two-band* distribution of surface states for Si(111). In spite of the importance of this face (the cleavage plane), the first realistic model calculations dealt with the (110) surface. This is not surprising, because the geometry of the diamond-like lattice is such that, for a fixed value of the component of the wave vector on the surface plane, the calculation for the (110) face is essentially reduced to a suitable one-dimensional model, which is not the case for the (111) face.

The first detailed calculations of localized states for Si(110), based on a realistic potential, were performed by Jones (1968). In this model, as in that suggested earlier by Chaves *et al.* (1966), the bulk crystal potential was unaltered up to the surface plane, where it changed abruptly to the vacuum level, which was determined by work-function measurements. Using the pseudopotential (NFE) method, Jones obtained *two bands* of surface states. Subsequently, Heine and Jones (1969) developed a theory for the band structure of diamond-type semiconductors, which provided a simple scheme for calculating the bulk-band energy levels. This scheme, in combination with the standard matching method, was used by Yndurain and Elices (1972) and Jones (1972) for surface-state calculations on Si(110) and Si(111). For the (110) face, both papers presented similar results, i.e., *two bands* of surface states were found, in good agreement with Jones' earlier investigations. However, for Si(111), the results were rather different, since Yndurain and Elices obtained *three bands* of surface states, while Jones found only two. Although the reason for such a difference was not clear, Jones' finding was in qualitative agreement with the results of Alstrup (1970, 1971) and Bortolani *et al.* (1971).

Alstrup (1970, 1971) set up a realistic LCAO (tight-binding) model for finite semiconductor crystals, using the parameters of Dresselhaus and Dresselhaus (1967), which reproduced the bulk bands much more accurately, and calculated the surface bands of Si(111) and (110) slabs. The results suggested that only small differences exist between the surface-state distribution of these two surfaces. *Two bands* of surface states were also obtained by Bortolani *et al.* (1971), who adopted the Green-function approach to Penn's (1962) model. The Penn model is based on the approximation that an isotropic FEG separates the filled from the empty states, and gives a fairly good description of the energy-band structure of infinite semiconductors (Heine and Jones 1969). Thus, the two-band model seemed to be theoretically established. However, Appelbaum and Hamann (1973) reported the first self-consistent calculation of surface states for the Si(111) surface, and found only a *single band* lying totally in the FEG. Almost simultaneously, there appeared a paper by Flores and Rubio (1973). Within the framework of the Heine-Jones model, they analysed the matching at the (111) face of a diamond lattice and again obtained *two surface bands.* The situation was somewhat clarified by Elices *et al.* (1974), who used the surface Green-function method (Garcia-Moliner and Rubio 1969, 1971). The calculation yielded a *single band* of surface states, in agreement with the Appelbaum and Hamann results (for unrelaxed surfaces), but in sharp contrast to the other papers. However, Elices *et al.* carefully checked all the previous calculations and found a sign mistake in one of the equations of the paper by Flores and Rubio (1973). A correct evaluation led to only one band of states. They also pointed out that the two bands obtained by Alstrup (1970) correspond to a slab of finite width, while all other calculations were performed for a semi-infinite crystal with only one surface. Thus, only the limiting case of Alstrup's results can be compared with semi-infinite model calculations. However, as the width of the slab increases, the two bands obtained by Alstrup coalesce into a doubly degenerate band. Consequently, the four different calculations, using different methods, were essentially in agreement with the results of the *one-band* model.

9.3.3 Surface-state spectroscopy

Considerable progress in our understanding of surface states has been achieved, since the advent of optical techniques. A review of optical spectroscopy of surface states has been given by Heiland and Mönch (1973), Mönch (1973), Lüth (1975), and more recently by Eastman (1980).

Chiarotti *et al.* (1968, 1971) presented direct measurements of the optical absorption due to the presence of a band of surface states below the bottom of the conduction band on cleaved Ge (111) surfaces. The same technique of multiple internal reflection was also used later by Chiarotti *et al.* (1971) for determining the energy distribution of surface states for the Si(111) face. In this experiment, a beam of monochromatic light of intensity I_0 and energy $E < E_g$, the FEG width, was totally reflected several times on the (111) face, which had been freshly cleaved in ultra-high vacuum (Fig. 9.11). The intensity I of the light traversing the sample was recorded

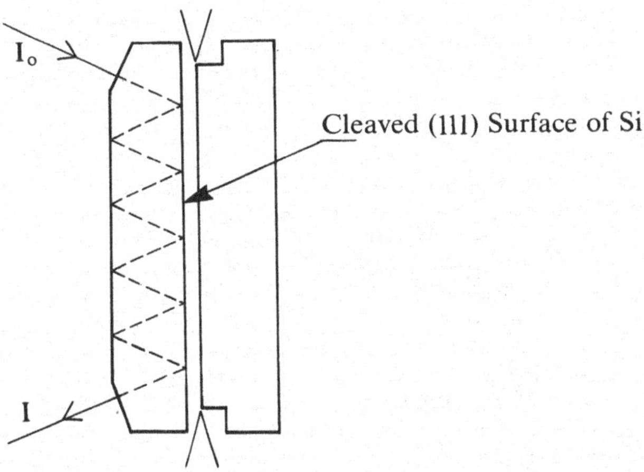

Fig. 9.11. Measurement of optical absorption of Si, after cleavage, via multiple internal reflection method. $I_0(I)$ is intensity of incident (multi-reflected) monochromatic light. (After Chiarotti *et al.* 1971.)

immediately after cleavage. Now, if the energy of the light $E < E_g$ (which means that it is not sufficient to excite the electrons from the valence to conduction band) and absorption (i.e., $I < I_0$) is still observed, then surface states must be responsible for the recorded optical absorption. Chiarotti *et al.* (1971) measured this absorption after cleavage (clean surface) and after complete oxidation of the surface. The results obtained were explained by the hypothesis that optical transitions take place between two bands of surface states, localized in the FEG, one band below and one above E_f at the surface. In other words, the Chiarotti *et al.* results strongly supported the *two-band-model* of surface states. Surface photoconductivity measurements (Müller and Mönch 1971) also led to results in close agreement with those of the Chiarotti *et al.* two-band model (cf. Fig. 9.12a). However, this picture was challenged by surface photoemission experiments, in which ultraviolet light was shone on the sample and the energy distribution of the photoemitted electrons measured. Such measurements were performed by Eastman and Grobman (1972), Wagner and Spicer (1972) and Erbudak

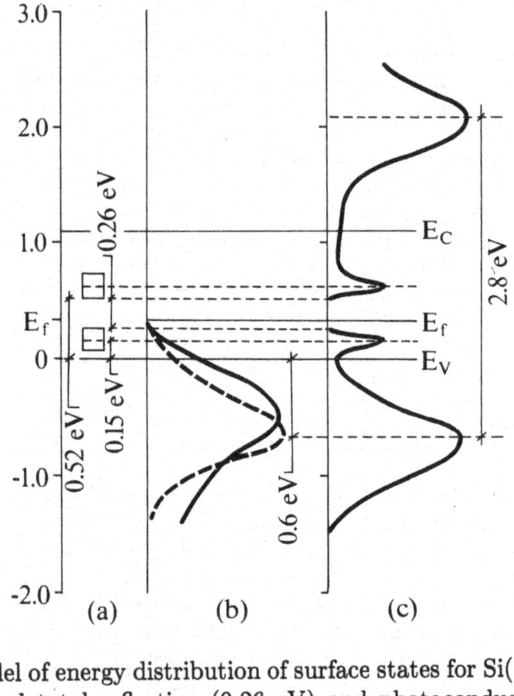

Fig. 9.12. Model of energy distribution of surface states for Si(111) constructed from: (a) internal total reflection (0.26 eV) and photoconductivity (0.52 eV) measurements, (b) photoemission experiments (solid line), after Eastman and Grobman (1972); (broken line) after Wagner and Spicer (1972), (c) a combination of the existing experimental data and new ellipsometric (2.8 eV) measurements (Meyer and Kroes 1975.)

and Fischer (1972). Although the results obtained by them did not quite coincide, they indicated that surface states form *one band* with a main peak approximately 0.6 eV below the top of the valence band (cf. Fig. 9.12b). Evidence for the existence of surface states well into the valence band was also given by electron energy-loss experiments (Rowe and Ibach 1973).

Meyer and Kroes (1975) pointed out that ultraviolet photoemission spectroscopy gives information primarily about occupied surface states. Thus, in order to obtain a complete picture of the surface-state distribution, data are needed from both the photoemission method, to derive the energy distribution of the occupied surface states, and also from such methods as electron energy-loss spectroscopy or ellipsometry (Meyer *et al.* 1971, Meyer 1974, Morgan 1974) to obtain information on the empty levels. On the basis of the existing experimental data and ellipsometric measurements, Meyer and Kroes constructed the *two-band model* of surface states shown in Fig. 9.12c.

Photoemission spectroscopy and, in particular, angle-resolvent techniques (Eastman 1980, Himpsel 1983) were later successfully applied to semiconductor surfaces and were used to determine many features of the surface electronic structure. Experimental results of this period, obtained for Si(100) and Si(111) surfaces, were summarized by Eastman (1980): 'angle-resolved and angle-integrated photoemission studies have observed that Si(111)-(2 × 1) is semiconducting with an occupied 'dangling bond' surface band near E_v, while optical studies (Chiarotti *et al.* 1971) show the surface to be semiconducting with a surface-state optical gap $\sim$ 0.3 eV'.

Thus, at the beginning of the 1980's it was rather difficult to construct a consistent picture of the energy distribution of surface states – the photoemission results, strongly supported by theoretical considerations, indicated the presence of one band of surface states, whereas, from the optical absorption measurements, it followed that two surface bands were formed. However, in all the theoretical approaches, the crystallographic bulk structure was assumed to be periodic up to the surface, i.e., no atomic rearrangement of the surface (surface reconstruction), as follows from LEED results (Eastman 1980), was taken into account. This is not surprising, since theoretical investigations of the correlation between surface states and surface structure are as interesting and important as they are difficult to perform. The complex surface geometry gives rise to complicated boundary conditions, so, any attempt to consider a crystal surface with rearrangement requires rather formidable mathematical analysis. In the late 1950's and 1960's, this difficulty was avoided by utilizing simple crystal models. However, since the 1970's, the use of large computers has made it possible to perform calculations of the electronic structure of reconstructed semiconductor surfaces.

9.3.4 Surface reconstructed crystals

The first attempt to take account of the extent of the surface perturbed region was made by Koutecký (1957). Employing the one-electron resolvent method, in conjunction with the TBA, he studied the effect of surface potential penetration on the surface states by perturbing the Coulomb integrals of the first $(Z + 1)$ cells. The results obtained showed that, when certain existence conditions were satisfied, it was possible for $(Z+1)$ surface states to occur. For $Z = 1$, two surface states with different energies were found to exist. The wave function amplitude of the higher-energy state decreased monotonically into the crystal, while the amplitude of the lower one attained its maximum in the second cell beneath the crystal surface. Such states he called *subsurface* states.

By way of a confirmation of the above findings, Phariseau (1960) investigated the same problem via a GF solution of the Schrödinger equation for a perturbed KP model with δ-wells. Contrasting this approach with the previous orbital one, shows that the depths of the Dirac δ - potential wells and the lattice spacing are analogous to the Coulomb and resonance integrals, respectively. However, instead of representing the surface distortion by modifying the depth of the first $(Z+1)$ δ-potential wells, Phariseau chose to modify the lattice spacings. Nevertheless, his results were in agreement with those of Koutecký (1957). Thus, it can be concluded that: (a) the results are independent of the model, and (b) the appearance of surface states is determined solely by the presence of the perturbation, whether it influences the δ-potential strength (Coulomb integrals) or the lattice spacing (resonance integral).

Since, in any real crystal, the surface distortion affects both the atomic potential and bond length, a natural extension of Koutecký's and Phariseau's work was to adopt a model which combines the surface deformation features of both their treatments (Stęślicka and Wojciechowski 1966). For simplicity, the surface deformation was confined to the immediate vicinity of the crystal surface. A more detailed study (Stęślicka 1966) of the surface-state existence condition and energy equation revealed that from zero to two surface states could exist, depending on the values of the various crystal model parameters. The existence conditions of surface states were also investigated in a three-dimensional deformed (the displacement of the surface atoms in the plane normal to the surface) crystal (Stęślicka 1970). Using the direct crystal orbital method, within the framework of the TBA and including overlap, it was shown, by utilizing the 2*s*-state wave function of Duncanson and Coulson (1944), that a decrease of about 8% in the surface lattice constant was sufficient to produce an additional band of surface states, i.e., two surface bands can exist in the energy gap. In other work, both normal and lateral distortions of the (110) plane of zinc-blende surface atoms were investigated by Levine and Freeman (1970), using the resolvent crystal orbital - TBA method (Koutecký 1957, Freeman 1970) and earlier one-dimensional treatments of the model (Davison and Koutecký 1966, Levine and Davison 1968). Surface deformation of binary crystals was also studied during this period (Davison 1967).

The first self-consistent calculation of surface states was reported by Appelbaum and Hamann (1973), who constructed realistic potentials for the Si(111) surface with the atoms in their ideal lattice positions and with the last atomic plane relaxed in the normal direction by 0.33 Å. The results were based on a method previously introduced (Appelbaum and Hamann 1972) for solving the Schrödinger equation exactly for any potential in the last few atomic layers and the nearby vacuum. For an ideal structure, a single band of surface states was found, lying in the FEG between the conduction and valence bands. For the relaxed geometry, however, *three* bands of surface states were obtained – the topmost one lying in the FEG; the second and third overlapping the bulk valence band. These results were confirmed by Pandey and Phillips (1974), using the semi-empirical TB approach, and by the more recent work of Schlüter *et al.* (1975), who used slab geometry. The ideal and relaxed Si(111) surface was also investigated by Ciraci *et al.* (1975), using the bond-orbital model (Harrison and Ciraci 1974).

The calculated energy-band structure of surface states is shown in Fig. 9.13. For an ideal surface, a single, half-filled band of total width 1.4 eV was found. Contrary to previous calculations, this band overlaps both the valence band and the FEG (see Fig. 9.13, state S_1), in reasonable agreement with photoemission measurements. A detailed analysis has also shown another band of surface states at the bottom of the valence band (Fig. 9.13, state S_2). The total density of states of the relaxed Si(111) surface is depicted in Fig. 9.13 by the broken lines. In accord with previous results, three bands of surface states were obtained. However, the upper band still overlapped the valence band, although the energy corresponding to the state S_1 was higher in the FEG.

When the Si(111) surface is prepared by cleavage, it occurs in a metastable (2 × 1) form, in which atoms of the outermost atomic layer are alternatively moved inward and outward (i.e., buckling) to form a (2 × 1) planar unit cell (Haneman 1968, Taloni and Haneman 1968, Schlüter *et al.* 1975). Consequently, calculation of this reconstructed surface is more meaningful for comparison with experimental data.

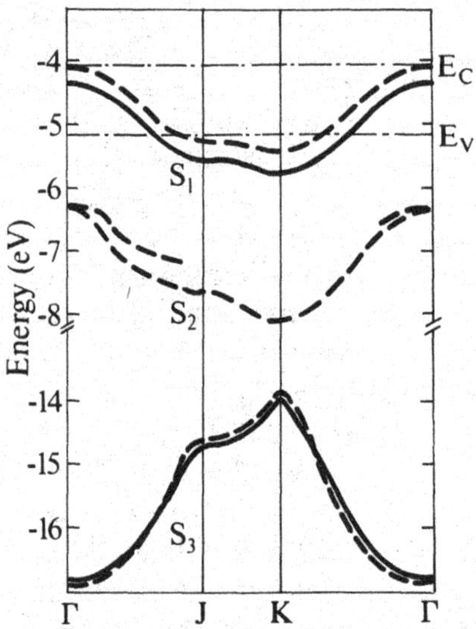

Fig. 9.13. Energy-band structure of surface states for ideal (solid line) and relaxed (broken lines) Si(111). (After Ciraci *et al.* 1975.)

Betteridge and Heine (1974) accepted a single-band model, for the ideal Si(111) surface, and incorporated into the model a FEG, arising from the surface reconstruction into a (2 × 1) superlattice structure. For such a model, a detailed calculation of the optical absorption between surface states was reported and the results were compared with the observed absorption (Fig. 9.14). The good agreement (in both shape and absolute magnitude) between the calculated and experimental curves strongly supports the hypothesis of Chiarotti *et al.* (1971) that the measured absorption arises from surface states.

The electronic structure of the (2 × 1) reconstructed Si(111) surface has been the subject of a large number of theoretical calculations (Schlüter *et al.* 1975, Pandey and Phillips 1975, Ciraci *et al.* 1975, Appelbaum and Hamann 1975). The first three papers, which appeared almost simultaneously, reported that, upon reconstruction, the state S_1 (Fig. 9.13) splits into two states S_1' and S_1'' (Fig. 9.16). Ciraci *et al.* (1975), however, found that the states S_1' and S_1'', separated by about 0.5 eV, broaden into two half-occupied overlapping bands and thus, 'the reconstructed surface is perhaps metallic, or a narrow-band-gap semiconductor, depending on the reconstruction parameters.' Schlüter *et al.* (1975) calculated the DOS for

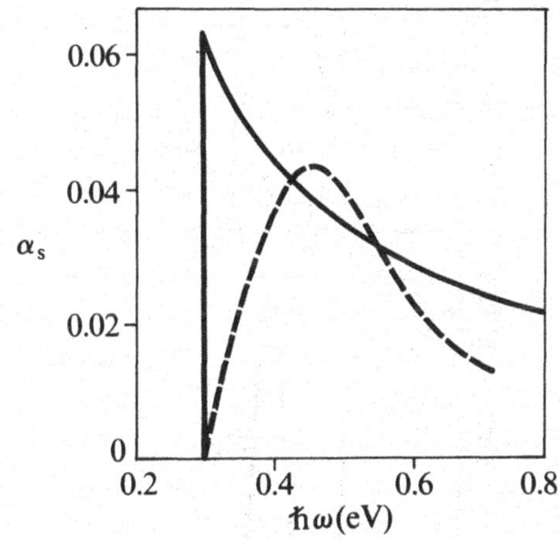

Fig. 9.14. Calculated surface-state absorption for simulated (2 × 1) reconstructed Si(111) (full line), compared with observed absorption (broken line). (After Betteridge and Heine 1974.)

a twelve-layer Si(111) slab (Fig. 9.15). For the ideal surface, one very flat surface band was found at about mid-gap, which was almost dispersionless and half-occupied, placing E_f right at the peak (Fig. 9.15a). On reconstruction, this band splits, exhibiting two peaks (Fig. 9.15b), corresponding to *two* surface bands separated by a FEG. However, if this FEG is smaller, or comparable to the dispersion of the surface bands, it may not lead to a vanishing DOS between bands (Fig. 9.15). Two separate bands of surface states, one of which was occupied, allowed for infrared absorption, as observed by Chiarotti *et al.* (1971) (cf. Fig. 9.12a), and were in agreement with the theoretical calculation of Betteridge and Heine (1974).

In addition to the appearance of two split surface bands, Schlüter *et al.* (1975) also observed a shift towards lower energies, placing the occupied surface state in the vicinity of the upper limit of the valence band (Fig. 9.15), in reasonable agreement with the photoemission measurements (Fig. 9.12b).

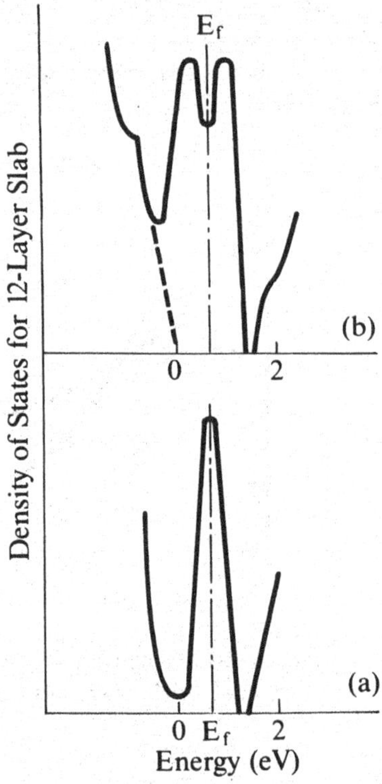

Fig. 9.15. Calculated density of states for 12-layer Si(111) slab for (a) ideal and (b) reconstructed surfaces. (After Schlüter *et al.* 1975.)

Pandey and Phillips (1975) have also studied the (2×1) reconstructed Si(111) surface, using empirical tight-binding calculations, and obtained the splitting of the surface band. A detailed study of the electronic structure of this surface was presented by Ciraci *et al.* (1975), and the calculated total DOS is shown in Fig. 9.16. The characteristic splitting of S_1 into two peaks S_1' and S_1' is again observed. The back-bonding state S_2 (cf. Fig. 9.13) near the top of the valence band splits into four resonating states B_1, B_2, B_3 and B_4. Also, the state S_3 near the bottom of the valence band splits into two distinct bands S_3' and S_3''. The surface states R_2, R_1 and T_1 are the so-called *regional surface states*, which are found in some special regions of the surface Brillouin zone. Since their charge is localized in deeper layers, they are not affected by reconstruction (cf. also Chadi and Cohen 1975, Joannopoulos and Cohen 1975, Schlüter *et al.* 1975).

Although there were still some discrepancies between the results of different authors, it appeared to be generally accepted that, for an ideal surface, only one band of surface states can exist, while reconstruction causes the splitting of this band into two bands. It is worth noting that such conclusions also follow from simple modelistic approaches.

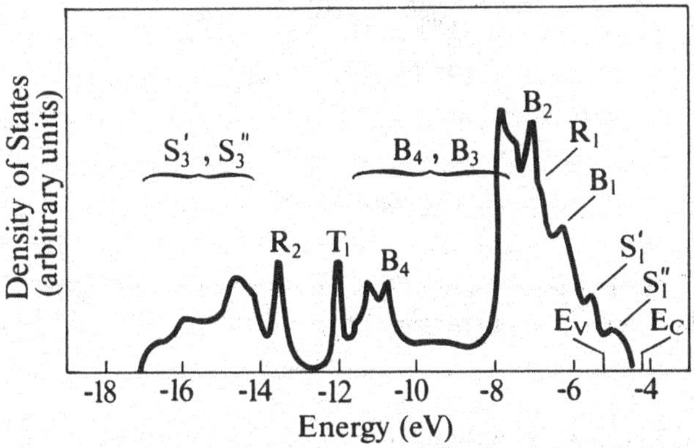

Fig. 9.16. Calculated total density of states for (2 × 1) reconstructed Si(111). (After Ciraci *et al.* 1975.)

In 1981, Pendry published his theory of inverse photoemission, and contrasted it with the angle-resolved ultraviolet photoemission technique. Inverse photoemission was soon heralded as one of the most promising new developments in surface science, since it provided a very powerful means of studying unfilled (empty) states of a solid, by observing the radiative transitions of an incoming electron, in a free-electron-like initial state, into an empty final state above E_f.

The momentum of the incident electron parallel to the surface is conserved in the radiative transition, since the momentum of the emitted ultraviolet photon is negligible. Knowing the energies and momenta of the

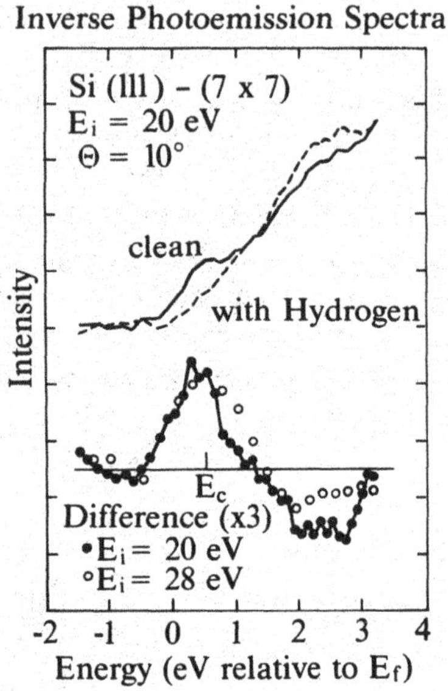

Fig. 9.17. Inverse photoemission spectra for clean (solid line) and hydrogen-covered (broken line) (7 × 7) reconstructed Si(111). (After Fauster and Himpsel 1983.)

final states, it is possible to determine the unfilled energy levels of a crystal.

The first momentum-resolved inverse photoemission experiment[2] was reported by Fauster and Himpsel (1983), who observed the empty surface states of the annealed Si(111)-(7 × 7) surface. Fig. 9.17 shows inverse photoemission spectra of this surface before and after exposure to hydrogen. The difference spectra in the lower part of Fig. 9.17 clearly show empty states below the conduction band minimum $E_c = E_f + 0.53$ eV (Himpsel *et al.* 1983), extending down to E_f. This finding was consistent with the

[2]Earlier experiments (Denninger *et al.* 1982, Woodruff and Smith 1982) were limited by a fixed photon energy and a resolution of 0.7 eV.

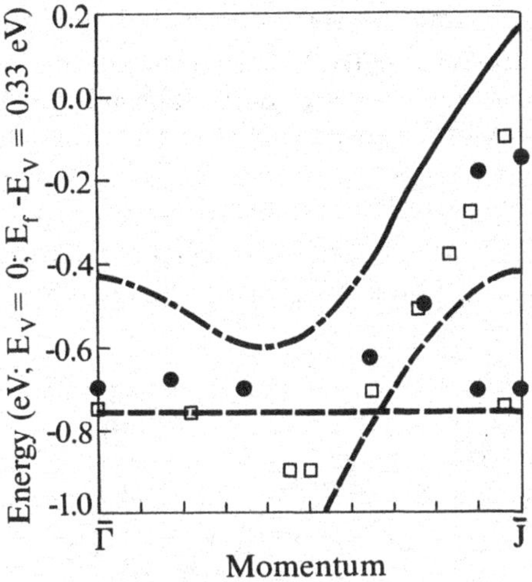

Fig. 9.18. Comparison of experimental surface state dispersions with theoretical band dispersions along the long axis $\overline{\Gamma J}$ of the surface Brillouin zone of Si(111)-(2 × 1). Full circles are from Himpsel *et al.* and open squares are from Uhrberg *et al.* (1982) using peak positions of angle-resolved photoelectron spectra. The full curve is a self-consistent band calculation for Pandey's π-bonded chain model, where the dot-dash section is degenerate with bulk states. The two dashed curves are for a correlated cluster calculation of a relaxed antiferromagnetic surface by Redondo *et al.* (1982). Note that all data points could shift upwards by up to 0.15 eV using recent values for the Fermi level position. (After Himpsel *et al.* 1983.)

metallic behaviour of the Si(111)-(7 × 7) surface determined via photoemission (Himpsel *et al.* 1981). Thus, the two-band model of surface states seemed to be well-established, though there were still different interpretations of the angle-resolved photoemission data for Si(111)-(2×1), regarding the lower (occupied) band of surface states. A comparison of experimental surface-state dispersions with theoretical band dispersions along the $\overline{\Gamma J}$ axis of the surface Brillouin zone for Si(111)-(2 × 1) is given in Fig. 9.18, while Fig. 9.19 displays two results for the surface state at the $\bar{J}$ point. As can be seen from Fig. 9.19, photoemission spectra exhibit two peaks (cf. Rowe *et al.* 1974, 1975). The surface-state interpretation of these peaks (Himpsel *et al.* 1981) was triggered by a correlated-cluster calculation (Redondo *et al.* 1982), which predicts two surface-state bands (Fig. 9.18). On the other hand, the band-like states, expected from the π-bonded chain model, give a qualitatively correct dispersion curve (Fig. 9.18), but cannot explain a second surface state at $\bar{J}$. The puzzle was finally resolved by Uhrberg *et al.* (1984), who showed that the dominant structure in the photoemission spectra is due to a direct transition from the uppermost valence band to a final band with free-electron-like dispersion. Consequently, this structure should not be considered, when comparing experimental and calculated surface-state dispersions for the Si(111)-(2 × 1) surface.

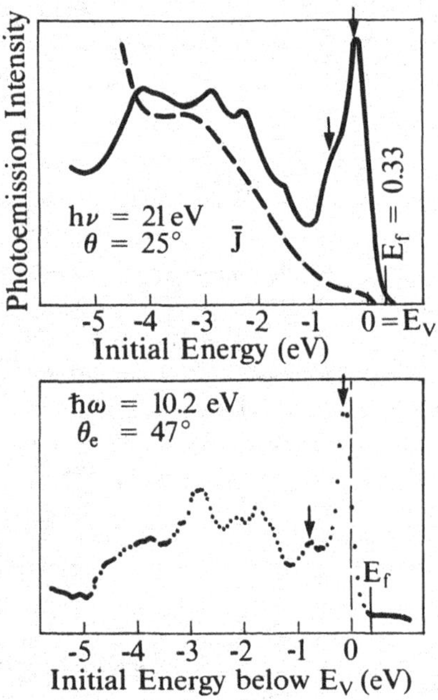

Fig. 9.19. Comparison of two data sets for the surface state at the $\bar{J}$ point of Si(111)-(2×1). Two surface states (ticmarks) are seen with similar relative areas. The dashed line is for a H-covered surface where surface states are quenched. (After Himpsel *et al.* 1983.)

An important step towards a better understanding of a wide variety of spectroscopic data on surface states was made, when Pandey (1981, 1983) reported his theory of semiconductor surface reconstruction for Si(111)-

(7 × 7) and Si(111)-(2 × 1), and proposed the π-bonded chain model (Fig. 9.20). In this model, the dangling bonds, which on the ideal surface are placed on the second-neighbour atoms, are brought to the nearest neighbours to form a one-dimensional zig-zag chain. The increased overlap of the dangling bonds, as well as the formation of a one-dimensional energy band, permitted by this geometry, are responsible for the large energy benefit and the width and dispersion of the dangling-bond surface band. Contrary to other models, the π-bonded model was found to be consistent with all the experimental data – an illustration of this agreement is shown in Fig. 9.21. Calculations performed for diamond (Pandey 1982) and Ge(111)-(2 × 1) (Pandey 1982, Northrup and Cohen 1983) surfaces extended the validity of the π-bonded model to those surfaces as well. According to these calculations, one filled bonding surface-state band is formed on Ge(111)-2 × 1, as well as one empty antibonding band.

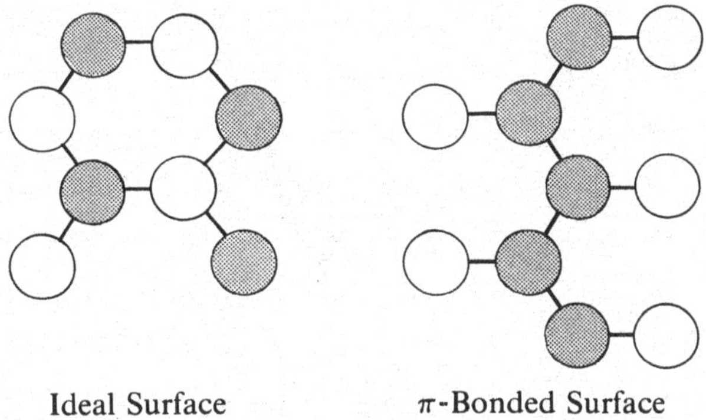

Fig. 9.20. Geometry of the ideal Si(111) surface and its modification by π-bonding reconstruction. Open circles represent second layer atoms that do not have any broken bonds. The shaded circles represent surface atoms. Each surface atom has one broken bond that points out of the surface. The π-bonding reconstruction brings the broken bonds to the nearest-neighbour atoms. (After Pandey 1983.)

In 1985, Nicolls *et al.* performed the first direct photoemission study of the surface-state band on Ge(111)-(2 × 1), which is empty for a neutral surface, but, in an n-doped crystal, is partially filled to provide the surface charge for band bending and Fermi-level pinning. The results of this experiment proved the existence of a new surface-state structure close to

E_f, in good accord with the calculated antibonding band of the π-bonded chain model. The obtained value for the surface-state band FEG (0.5 eV) also agrees with the optical absorption measurements of the FEG.

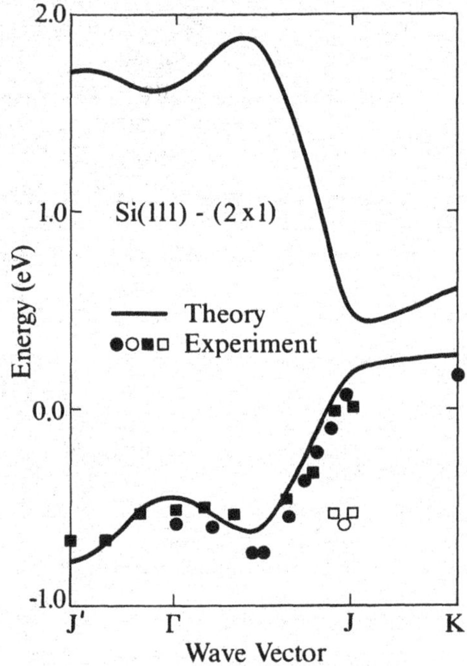

Fig. 9.21. Surface-state energy bands for the optimized π-bonded chain model of Si(111)-(2 × 1) compared with bands derived from angle resolved photoemission measurements. Two sets of measurements are shown: squares and circles. Filled symbols show the strong peaks in the spectra. The open symbols show weak features in the spectra, which depend on the cleavage process, and are not intrinsic to the surface. Energies are measured relative to the valence-band maximum. The Fermi energy is assumed to be at 0.5 eV. (After Pandey 1983.)

A similar agreement was also obtained in the first measurement of the polarization dependence of the Ge(111)-(2 × 1) surface-state absorption (Olmstead and Amer 1984). Consequently, their results strongly support the π-bonded chain model, and lead to the conclusion that 'similarity in the magnitude, dispersion and polarization dependence of the optical absorption by Si(111)-(2 × 1) and Ge(111)-(2 × 1) support the premise that the two reconstructions are similar.'

Therefore, both the experimental (direct and inverse photoemission and optical absorption) and theoretical results lead to a consistent *two-band* model of surface states for Ge(111) and Si(111). Finally, it is worth noting that self-consistent pseudopotential calculations (Pandey 1982) have shown that the generally accepted models of reconstruction, involving buckling, are still valid for heteropolar semiconductors, such as GaAs(100).

9.4 Surface States of Metals

In principle, surface states can exist in any crystal having free surfaces, i.e., not only in semiconductors, but also in metals and insulators. Theoretical predictions of metallic surface states were given in the first classic papers (Maue 1935, Goodwin 1939), but the practical problems of how they could be detected experimentally discouraged further investigations.

The first detailed study of the possible existence of surface states in NFE metals was given by Forstmann (1970). A NFE band structure exhibits two types of FEG, where the existence of surface states is possible. Since all previous investigations, with the exception of Bartoš (1969), were concerned with a FEG of type A at a Brillouin zone boundary, Forstmann studied a FEG of type B, resulting from the crossing of two bands inside the Brillouin zone. The result was that, in such a FEG, a surface state always exists. This encouraged Forstmann and Pendry (1970) to investigate the existence of surface states on *d*-band metals, since their band structure shows a FEG inside the Brillouin zone, due to the crossing of the *s*- and *d*-bands of the same symmetry. Using the wave-function matching method at the crystal surface, it was found that, for the (100) surface of a fcc transition metal, there exists a surface state at the lower end of the *d*-band region. For Cu(Ni) it lies $5 - 6$ eV($4 - 5$ eV) below E_f. But these are just the positions where many photoemission spectra show an additional peak in the DOS of Ni (Spicer 1966, Callcott and MacRae 1969) and Cu (Berglund 1966, Blodgett and Spicer 1965). Since this peak could not be explained on the basis of the DOS derived from the bulk-band structure, and was shown to be strongly dependent on surface conditions (Callcott and MacRae 1969), the existence of surface states was suggested as the reason for this anomaly in the photoemission DOS (see also Forstmann and Heine 1970).

These considerations led Plummer and Gadzuk (1970) to postulate that the observed structure at $E = -0.37$ eV in the W(100) energy distribution, obtained from field emission measurements, was a surface state. This structure was first observed by Swanson and Crouser (1966, 1967) and, in an attempt to explain its source, a long discussion took place (Gadzuk and Plummer 1973), but no satisfactory explanation was given. Plummer and Gadzuk (1970) compared an experimental enhancement factor[3] for the (100) plane of W with the relativistic energy bands for the (100) direction of W, calculated by Loucks (1965), and showing the two FEG created by the splitting of the three Δ_7 bands (Fig. 9.22). They suggested that the peak at $E = -0.37$ eV was a surface state in the spin-orbit split Δ_7 symmetry bands, and that the small peak at $E = -1.5$ eV reflected a high bulk DOS. The low-energy structure at $E \cong -1.5$ eV was also originally interpreted as a surface state (Plummer and Gadzuk 1970), but later experiments (Plummer and Bell 1972) made that interpretation rather questionable.

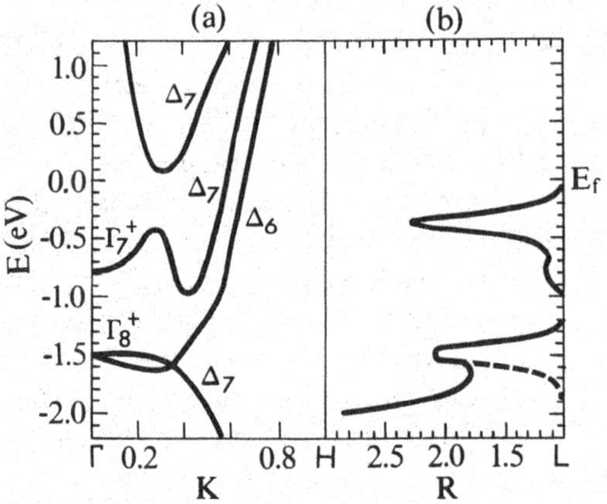

Fig. 9.22. (a) Loucks' (1965) calculated relativistic energy bands for (100) direction of W. (b) Experimental enhancement factor for W(100) (Plummer and Gadzuk 1970). (After Gadzuk and Plummer 1973.)

Using the results of Forstmann *et al.* (1970), Gadzuk (1972) constructed a theory of tunneling from surface states in metals in which the total surface-state current was given in terms of parameters obtainable from the bulk-band structure, work function, applied electric field and density of

[3]An enhancement factor $R(E)$ is obtained by dividing the measured energy distribution $j'(E)$ by a calculated free-electron energy distribution $j_0'(E)$, i.e., $R(E) = j'(E)/j_0'(E)$.

surface atoms. In the same year, however, Duke and Fauchier (1972) presented a model calculation, and suggested an explanation of the observed peak at $E = -0.37$ eV without the presence of surface states. All this discussion led Gadzuk and Plummer (1973) to the following conclusion: 'we cannot unequivocally demonstrate by the use of field emission energy distribution that we have observed a surface state.'

There then followed a detailed study of the band-structure effects on the field-emission energy distribution of W by Nicolaou and Modinos (1975, 1976), which were also discussed in Modinos' book (1984). The potential inside the metal was represented by the superposition of atomic potentials in the muffin-tin approximation, as evaluated by Mattheiss (1965), and terminated abruptly at the surface. The potential outside the crystal was represented by a triangular barrier. Numerical results of the total energy distribution of field-emitted electrons from the (100) and (110) planes of W were interpreted in terms of surface DOS effects. For the (100) plane, *virtual* surface states (Modinos 1974) were shown to exist in an energy region extending from 0.20 eV to 0.95 eV below the Fermi level. It was found, however, that only the states at 0.29 eV below the Fermi level gave rise to peaks in the total energy distribution of the emitted electrons. This strongly supports the suggestion of Plummer and Gadzuk (1970) that the 'mysterious' peak, experimentally observed at $E = -0.37$ eV, is due to the surface states. Nicolaou and Modinos (1976) replaced the triangular surface potential by the image-potential barrier, which shifted the band of surface states towards lower energies (Kolář and Bartoš 1973), so that the surface states lay between 0.56 eV and 0.85 eV below E_f. The peak in the total energy distribution of the emitted electrons was found near the upper limit of the band, where the DOS is maximum, i.e., around $E = -0.56$ eV. Thus, the calculated surface-resonance peak lies about 0.2 eV below the experimentally observed value, for the image-potential barrier, and about 0.1 eV above the experimental value for the triangular barrier. Since the real surface barrier lies somewhere between the two barriers, Nicolaou and Modinos concluded that 'the present theory is basically correct and that the exact position of the peak is sensitive to the choice of the surface barrier.'

However, there is still some arbitrariness in the above interpretation of the experimental data, which is connected with the possible existence of *true* surface states. Nicolaou and Modinos did not investigate in detail the true surface states, they only pointed out that such states are not likely to be important as far as field emission is concerned, although their contribution to the surface DOS can be significant.

General criteria for the existence of surface states at the centre and edge of the Brillouin zone were given by Pendry and Gurman (1975), and the results of a calculation of the surface-state energies in the $s-d$ hybridization FEG for Cu and W(100) surfaces were reported by Gurman (1976). Ab initio calculations on the (100), (110), and (111) surfaces of W and Mo were presented by Stephenson and Bullett (1984). The reasonable agreement between the distribution of surface states and experimental data led again to the conclusion that the structure found by Swanson and Crouser (1966, 1967) was the first observation of surface states on transition metals.

Although the role of surface states in the field-emission process was recognized and well-reviewed (Gadzuk and Plummer 1973, Modinos 1984), it is only within the last few years that a more detailed theoretical understanding of the surface-state contribution to the tunnelling current became available. (Davison and Kolář 1985, Kolář and Davison 1985, Davison *et al.* 1987).

9.5 Concluding Remarks

Appreciable progress has been made during the last two decades or so, in both theoretical and experimental studies of surface states. However, the mathematical difficulties arising from the boundary conditions at the crystal surface, and the consequent drastic approximations, as well as some arbitrariness in the interpretation of the experimental data, make our understanding of the basic properties of surface states still quite far from ideal. In 1972, Gadzuk characterized this situation in a very pictorial way: 'We are just left with the delightful metaphysical problem before us with regards to our god-like ability of creation or destruction within the realm of surface-state research.' Even today, we should still be aware of the possibility that some surface states can be created or destroyed by the very act of constructing theoretical models or performing measurements themselves.

And so, at last, we have come to the end of our journey, a journey which has taught us that *surfaces*, like all of Nature, are the manifestation of God to man, and that, as we grow to know them, more and more of their mysteries will be revealed to us. For, after all, life itself is but a journey to understanding. Since, as St. Paul said in his *First Epistle to the Corinthians*, 'When I was a child, I spake as a child, I understood as a child, I thought as a child: but when I became a man, I put away childish things. For now we see through the glass, darkly; but then *face* to *face*.'

APPENDICES

Sir, I have found you an argument,
I am not obliged to find you an understanding.
— Samuel Johnson

A. Projection Operators

If the wave function

$$|\psi^0\rangle = \sum_n c_n^0 |\phi_n^0\rangle \tag{A.1}$$

then

$$\begin{aligned}\langle\phi_m^0|\psi^0\rangle &= \sum_n c_n^0 \langle\phi_m^0|\phi_n^0\rangle \\ &= \sum_n c_n^0 \delta_{mn} = c_m^0 \end{aligned} \tag{A.2}$$

so (A.1) may be written as

$$|\psi^0\rangle = \sum_n |\phi_n^0\rangle\langle\phi_n^0|\psi^0\rangle \tag{A.3}$$

which shows that

$$\sum_n |\phi_n^0\rangle\langle\phi_n^0| = I \tag{A.4}$$

I being the *identity* operator of (8.18). Thus,

$$P_n^0 = |\phi_n^0\rangle\langle\phi_n^0| \tag{A.5}$$

is the *projection operator* for the subspace p_n^0 (Messiah 1962, Taylor 1970).

B. Direct Green-Function Calculation

The one-dimensional Schrödinger equation is

$$[\nabla_x^2 + E - V(x)]\psi(x) = 0 \tag{B.1}$$

and the corresponding GF equation is

$$[\nabla_x^2 + E - V(x)]G(x,x') = -\delta(x-x'). \tag{B.2}$$

If the boundary conditions are *unmixed*, so that each boundary condition involves only one point of the boundary and not both [e.g. $\psi(0)=0$ is unmixed, while $\psi(0)=\psi(1)$ is mixed], then

$$G(x,x') = \psi_1(x)g_1(x'), \quad x < x' \tag{B.3}$$

$$= \psi_2(x)g_2(x'), \quad x > x' \tag{B.4}$$

$\psi_{1,2}(x)$ being solutions of (B.1) and $g_{1,2}(x')$ arbitrary functions to be determined.

Choosing $\psi_1(x)$ to satisfy one boundary condition for $x < x'$ and $\psi_2(x)$ to satisfy the other boundary condition for $x > x'$, and introducing the *unit step function*

$$\begin{aligned} \theta(x-x') &= 1, \quad x > x' \\ &= 0, \quad x < x' \end{aligned} \tag{B.5}$$

the GF may be written as

$$G(x,x') = \psi_1(x)g_1(x')\theta(x'-x) + \psi_2(x)g_2(x')\theta(x-x') \tag{B.6}$$

which on differentiating leads to

$$\begin{aligned} G'(x,x') &= \psi_1'(x)g_1(x')\theta(x'-x) - \psi_1(x)g_1(x')\delta(x-x') \\ &\quad + \psi_2'(x)g_2(x')\theta(x-x') + \psi_2(x)g_2(x')\delta(x-x') \end{aligned} \tag{B.7}$$

where

$$\theta'(x-x') = \delta(x-x') \tag{B.8}$$

from (B.5). However, the $\delta(x-x')$ terms in (B.7) must be zero for $G''(x,x')$ to be independent of $\delta'(x-x')$ as in (B.2), so

$$\delta(x-x')[\psi_2(x)g_2(x') - \psi_1(x)g_1(x')] = 0$$

i.e.

$$\psi_2(x)g_2(x) - \psi_1(x)g_1(x) = 0. \tag{B.9}$$

Meanwhile, differentiation of (B.7) gives

$$\begin{aligned} G''(x,x') &= \psi_1''(x)g_1(x')\theta(x'-x) - \psi_1'(x)g_1(x')\delta(x'-x) \\ &\quad +\psi_2''(x)g_2(x')\theta(x-x') + \psi_2'(x)g_2(x')\delta(x-x') \end{aligned} \tag{B.10}$$

which, when compared with (B.2), shows that

$$\delta(x-x')[\psi_2'(x)g_2(x') - \psi_1'(x)g_1(x')] = -\delta(x-x')$$

i.e.

$$\psi_2'(x)g_2(x) - \psi_1'(x)g_1(x) = -1. \tag{B.11}$$

Solving (B.9) and (B.11) yields

$$g_1(x) = -\psi_2(x)/W[\psi_1(x), \psi_2(x)] \tag{B.12}$$

$$g_2(x) = -\psi_1(x)/W[\psi_1(x), \psi_2(x)] \tag{B.13}$$

where

$$W[\psi_1(x), \psi_2(x)] = \psi_1(x)\psi_2'(x) - \psi_1'(x)\psi_2(x) \tag{B.14}$$

is the Wronskian of $\psi_{1,2}(x)$. Substituting (B.12) and (B.13) in (B.6) gives (Friedman 1956, Lukes 1969)

$$G(x,x') = \frac{\psi_1(x)\psi_2(x')\theta(x'-x) + \psi_2(x)\psi_1(x')\theta(x-x')}{W[\psi_2(x'), \psi_1(x')]}. \tag{B.15}$$

C. Wronskian

The one-electron Schrödinger equation is

$$\psi''(x) + [E - V(x)]\psi(x) = 0. \tag{C.1}$$

Then, since

$$W(\psi_1, \psi_2) = \psi_1\psi_2' - \psi_1'\psi_2 \tag{C.2}$$

it follows that

$$dW/dx = \psi_1\psi_2'' - \psi_1''\psi_2 \tag{C.3}$$

which, on using (C.1), gives

$$dW/dx = (E-V)\psi_1\psi_2 - \psi_1(E-V)\psi_2 = 0. \tag{C.4}$$

Hence, the Wronskian is independent of its argument, i.e., is a constant.

BIBLIOGRAPHY

If I have seen further than other men, it is by standing on the shoulders of Giants.
— Isaac Newton

Abramowitz, M. and Stegun, I.A. (eds.) (1972). *Handbook of Mathematical Functions.* National Bureau of Standards, Washington, D.C.

Adnot, A. and Carrette, J.D. (1977). *Phys. Rev. Lett.* **38**, 1084.

Aerts, E. (1960a). *Physica* **26**, 1047.

_______ (1960b). In *Solid State Physics in Electronics and Telecommunications.* (eds. M. Désirant and J.L. Michels) Vol. 1. Academic Press, New York.

_______ (1960c). *Physica* **26**, 1057.

_______ (1960d). *Physica* **26**, 1063.

Allen, F.G. and Gobeli, G.W. (1962). *Phys. Rev.* **127**,150.

Alstrup, I. (1970). *Surf. Sci.* **20**, 335.

_______ (1971). *Phys. Stat. Sol. (b)* **45**, 209.

Anderson, P.W. (1961). *Phys. Rev.* **124**, 41.

_______ (1978). *Rev. Mod. Phys.* **50**, 191.

Antončik, E. (1959a). *J. Phys. Chem. Solids* **10**, 314.

_______ (1959b). *Czech. J. Phys.* **9**, 291.

_______ (1961). *J. Phys. Chem. Solids* **21**, 137.

Appelbaum, J.A. and Hamann, D.R. (1972). *Phys. Rev. B* **6**, 2166.

_______ (1973). *Phys. Rev. Lett.* **31**, 106.

_______ (1975). *Phys. Rev. B* **12**, 1410.

Arfken, G. (1985). *Mathematical Methods for Physicists.* Academic Press, San Diego, 3rd edn.

Artmann, K. (1952). *Z. Physik* **131**, 244.

Ashcroft, N.W. and Mermin, N.D. (1976). *Solid State Physics.* Holt, Reinhart and Winston, New York.

Aspnes, D.A. and Handler, P. (1966). *Surf. Sci.* **4**, 353.

Baldock, G.R. (1952). *Proc. Camb. Phil. Soc.* **48**, 457.

Bardeen, J. (1947). *Phys. Rev.* **71**, 717.

Barribeau, J.M., Carette, J.D., Jennings, P.J., and Jones, R.O. (1985). *Phys. Rev. B* **32**, 131.

Bartoš, I. (1969). *Surf. Sci.* **15**, 94.

_______ (1978). *Phys. Stat. Sol. (b)* **85**, K127.

Bartoš, I. and Davison, S.G. (1985). *Phys. Rev. B* **32**, 7863.

Bartoš, I. and Máca, F. (1976). *Czech. J. Phys. B* **26**, 619.

Bartoš, I. and Velický, B. (1974). *Czech. J. Phys. B* **24**, 981.

Bartynski, R.A., Gustafsson, T., and Soven, P. (1985). *Phys. Rev. B* **31**, 4745.

Berglund, C.N. (1966). In *Colloquium on the Optical Properties and Electronic Structure of Metals and Alloys, Paris (1965).* (ed. F.A. Abelés). North-Holland, Amsterdam.

Betteridge, G.P. and Heine, V. (1974). *Surf. Sci.* **43**, 417.

Bloch, F. (1928). *Z. Physik* **52**, 555.

Blodgett, Jr., A.J. and Spicer, W.E. (1965). *Phys. Rev. Lett.* **15**, 29.

Bolotovskii, B.M. and Frenkel, V. (eds.). (1990). *I.E. Tamm: Selected Papers.* Springer-Verlag, New York.

Born, M. and von Kármán, Th. (1912). *Physik. Z.* **13**, 297.

_______ (1913). *Physik. Z.* **14**, 15.

Born, M. and Oppenheimer, J.R. (1927). *Ann. Physik* **84**, 457.

Bortolani, V., Calandra, C., and Sghedoni, A. (1971). *Phys. Lett.* **34A**, 193.

Brattain, W.H. and Shockley, W. (1947). *Phys. Rev.* **72**, 345.

Brillouin, L. (1930). *J. Phys. Radium* **1**, 377.

_______ (1953). *Wave Propagation in Periodic Structures.* Dover, New York.

Brust, D. (1964). *Phys. Rev.* **134**, 1337.

Callcott, T.A. and MacRae, A.U. (1969). *Phys. Rev.* **178**, 966.

Carver, T.R. (1971). *Am. J. Phys.* **39**, 1225.

Chadi, D.J. and Cohen, M.L. (1975). *Phys. Rev. B* **11**, 732; *Solid State Comm.* **16**, 691.

Chaves, C.M., Majlis, N., and Cardona, M. (1966). *Solid State Comm.* **4**, 271.

Chiarotti, G., Del Signore, G., and Nannarone, S. (1968). *Phys. Rev. Lett.* **21**, 1170.

Chiarotti, G., Nannarone, S., Pastore, R., and Chiaradia, P. (1971). *Phys. Rev. B* **4**, 3398.

Cho, H.S. and Prucnal, P.L. (1988). *Phys. Rev. B* **38**, 2063.

Ciraci, S., Batra, I.P., and Tiller, W.A. (1975). *Phys. Rev. B* **12**, 5811.

Cole, M.W. and Cohen, M.H. (1969). *Phys. Rev. Lett.* **23**, 1238.

Coulson, C.A. (1961). *Valence.* Oxford University Press, London.

Cutler, P.H. and Davis, J.C. (1964). *Surf. Sci.* **1**, 194.

Davison, S.G. (1967). *Physica* **37**, 539.

Davison, S.G. and Amos, A.T. (1965). *J. Chem. Phys.* **43**, 2223.

Davison, S.G. and Kolář, M. (1985). *J. Phys.* C**18**, 4581.

Davison, S.G. and Koutecký, J. (1966). *Proc. Phys. Soc.* **89**, 237.

Davison, S.G., Lavis, D.A., and Sulston, K.W. (1986). *J. Phys. Chem.* **90**, 652.

Davison, S.G. and Levine, J.D. (1970). In *Solid State Physics.* (eds. H. Ehrenreich, F. Seitz and D. Turnbull). Vol. 25. Academic Press, New York. Russian translation by A.Y. Belenky. (ed. D.A. Kirzhnits). Mir, Moscow, 1973; in commemoration of Igor Tamm's death in 1971.

Davison, S.G., Ling, T.Y., and Ghosh, U.S. (1967). *J. Phys. Chem. Solids* **28**, 1921.

Davison, S.G. and Stęślicka, M. (1969). *J. Phys. C* **2**, 1802.

Davison, S.G. and Tan, K.P. (1971). *Surf. Sci.* **27**, 297.

Davison, S.G., Ueba, H., and Jerrard, R.J. (1980). *J. Phys. C* **13**, 1351.

Davison, S.G., Ueba, H., Jerrard, R.J., Kolář, M., and Stęślicka , M. (1987). *Prog. Surf. Sci.* **25**, 97.

Davydov, A.S. (1968). *Quantum Mechanics.* Pergamon Press, Oxford.

Denninger, G., Dose, V., and Bonzel, H.P. (1982). *Phys. Rev. Lett.* **48**, 279.

Detcheva, V., Metodieva, M., Kandilarov, B.D., and Primatarowa, M.T. (1990). *Helv. Phys. Acta* **63**, 908.

Dietz, R.E., McRae, E.G., and Campbell, R.L. (1980). *Phys. Rev. Lett.* **45**, 1280.

Dirac, P.A.M. (1958). *The Principles of Quantum Mechanics.* Oxford University Press, London.

Dose, V., Altmann, W., Goldmann, A., Kolac, U., and Rogozik, J. (1984). *Phys. Rev. Lett.* **52**, 1919.

Dresselhaus, G. and Dresselhaus, M.S. (1967). *Phys. Rev.* **160**, 649.

Duke, C.B. (1969). *Tunnelling in Solids.* (Solid State Physics, Supp. 10). Academic Press, New York.

Duke, C.B. and Fauchier, J. (1972). *Surf. Sci.* **32**, 175.

Duncanson, W.E. and Coulson, C.A. (1944). *Proc. Roy. Sci. (Edinburgh)* **62A**, 37.

Eastman, D.E. and Grobman, W.D. (1972). *Phys. Rev. Lett.* **28**, 1378.

Eastman, D.E. (1980). *J. Vac. Sci. Tech.* **17**, 492.

Echenique, P.M. (1976). Ph.D. thesis, University of Cambridge, England.

Echenique, P.M. and Pendry, J.B. (1978). *J. Phys. C* **11**, 2065.

_______ (1989). *Prog. Surf. Sci.* **32**, 111.

Elices, M., Flores, F., Louis, E., and Rubio, J. (1974). *J. Phys. C* **7**, 3020.

Erbudak, M. and Fischer, T.E. (1972). *Phys. Rev. Lett.* **29**, 732.

Esaki, L. (1984). *J. Physique* **45**, C5-3.

_______ (1986). *IEEE J. Quantum Electron.* **22**, 1611.

Fairbairn, W.M., Glasser, M.L., and Stęślicka, M. (1973). *Surf. Sci.* **36**, 462.

Falicov, L.M. and Yndurain, F. (1975). *J. Phys. C* **8**, 147, 1563, 1571.

Fauster, T. and Himpsel, F.J. (1983). *J. Vac. Sci. Tech. A* **1**, 1111.

Feuer, P. (1952). *Phys. Rev.* **88**, 92.

Floquet, G. (1883). *Ann. Sci. Ec. Norm.* Paris sup.2, **12**, 47.

Flores, F., Garcia-Moliner, F., and Rubio, J. (1970). *Solid State Comm.* **8**, 1065.

Flores, F. and Rubio, J. (1973). *J. Phys. C* **6**, L258.

Foo, E.-N., and Wong, H.-S. (1974). *Phys. Rev. B* **9**, 1857.

Forstmann, F. (1970). *Z. Physik* **235**, 69.

_______ (1978). *Photoemission and the Electronic Properties of Surfaces.* (eds. B. Feuerbacher, B. Fitton and R.F. Willis). Wiley, New York.

Forstmann, F. and Heine, V. (1970). *Phys. Rev. Lett.* **24**, 1419.

Forstmann, F. and Pendry, J.B. (1970). *Z. Physik* **235**, 75.

Fowler, R.H. (1933). *Proc. Roy. Soc. A* **141**, 56.

Freeman, S. (1970). *Phys. Rev. B* **2**, 3272.

Friedman, B. (1956). *Principles and Techniques of Applied Mathematics.* Wiley, New York.

Gadzuk, J.W. (1972). *J. Vac. Sci. Tech.* **9**, 591.

Gadzuk. J.W. and Plummer, E.W. (1973). *Rev. Mod. Phys.* **45**, 487.

Garcia-Moliner, F. and Rubio, J. (1969). *J. Phys. C* **2**, 1789.

_______ (1971). *Proc. Roy. Soc. A* **324**, 257.

Garcia, N. and Solana, J. (1976). *Physica B* **81**, 296.

Garcia, N., Riehl, B., Frank, K.H., and Williams, A.R. (1985). *Phys. Rev. Lett.* **54**, 591.

Gay, J.G., Smith, J.R., and Arlinghaus, F.J. (1979). *Phys. Rev. Lett.* **42**, 332.

Giesen, K., Hage, F., Himpsel, F.J., Riess, H.J., and Steinmann, W. (1985). Phys. Rev. Lett. **55**, 300.

_______ (1986). *Phys. Rev. B* **33**, 5241.

_______ (1987). *Phys. Rev. B* **35**, 971.

Glasser, M.L. and Davison, S.G. (1970). *Int. J. Quantum Chem.* **3S**, 867.

Glasser, M.L. (1977). *Surf. Sci.* **64**, 141.

Gobeli, G.W. and Allen, F.G. (1962). *Phys. Rev.* **127**, 141.

Goodwin, E.T. (1939a). *Proc. Camb. Phil. Soc.* **35**, 205.

_______ (1939b). *Proc. Camb. Phil. Soc.* **35**, 221.

_______ (1939c). *Proc. Camb. Phil. Soc.* **35**, 232.

Grimley, T.B. (1960). *Adv. Catalysis* **12**, 1.

Gurman, S.J. (1976). *J. Phys. C.* **9**, L609.

Haken, H. (1976). *Quantum Field Theory of Solids.* North-Holland, Amsterdam.

Handler, P. (1963). *Appl. Phys. Lett.* **3**, 96.

Haneman, D. (1968). *Phys. Rev.* **170**, 705.

Hansson, G.V. and Flodström, S.A. (1978). *Phys. Rev. B* **17**, 473.

Harrison, W.A. and Ciraci, S. (1974). *Phys. Rev. B* **10**, 1516.

Heiland, G. and Mönch, W. (1973). *Surf. Sci.* **37**, 30.

Heimann, P., Miosga, H., and Neddermeyer, H. (1979a). *Phys. Rev. Lett.* **42**, 801.

Heimann, P., Hermanson, J., and Miosga, H. (1979b). *Phys. Rev. B* **20**, 3059.

Heimann, P., Hermanson, J., Miosga, G., and Neddermayer, H. (1979c). *Phys. Rev. Lett.* **43**, 1757.

Heine, V. (1963). *Proc. Phys. Soc.* **81**, 300.

_______ (1964). *Surf. Sci.* **2**, 1.

_______ (1965). *Phys. Rev.* **138**, A1689.

Heine, V. and Jones, R.O. (1969). *J. Phys. C* **2**, 719.

Hilderbrand, F.B. (1968). *Finite Difference Equations and Simulations.* Prentice-Hall, Engelwood Cliffs.

Himpsel, F.J. (1983). *Adv. Phys.* **32**, 1.

Himpsel, F.J., Eastman, D.E., Heimann, P., Reihl, B., White, C.W., and Zehner, D.M. (1981). *Phys. Rev. B* **24**, 1120.

Himpsel, F.J, Heimann, P., and Eastman, D.E. (1981). *Phys. Rev. B* **24**, 2003.

Himpsel, F.J., Fauster, Th., and Hollinger, G. (1983). *Surf. Sci.* **132**, 22.

Hoffmann, T.A. (1951). *Acta Phys. Hung.* **1**, 175.

_______ (1952a). *Acta Phys. Hung.* **2**, 97.

_______ (1952b). *Acta Phys. Hung.* **2**, 108.

_______ (1952c). *Acta Phys. Hung.* **2**, 195.

Hoffmann, T.A. and Kónya, A. (1951). *Acta Phys. Hung.* **1**, 5.

Hubbard, J. (1963). *Proc. Roy. Soc. A* **276**, 238.

Hückel, E. (1931a). *Z. Physik* **70**, 204.

_______ (1931b). *Z. Physik* **72**, 310.

Hulbert, S.L., Johnson, P.D., and Weinert, M. (1986). *Phys. Rev. B* **33**, 760.

Hussain, Z. and Smith, N.V. (1978). *Phys. Lett.* **66A**, 492.

Inglesfield, J.E. (1971). *J. Phys. C* **4**, L14.

Jahnke, E. and Emde, F. (1945). *Tables of Functions with Formulae and Curves.* Dover, New York.

James, H.M. (1949a). *Phys. Rev.* **76**, 1602.

_______ (1949b). *Phys. Rev.* **76**, 1611.

Jeffreys, H. and Jeffreys, B.S. (1962). *Methods of Mathematical Physics*, 3rd edn. Cambridge University Press, London.

Jennings, P.J. (1971). *Surf. Sci.* **25**, 513.

Jennings, P.J., Jones, R.O., and Weinert, M. (1988). *Phys. Rev. B* **37**, 6113.

Jennings, P.J. and Read, M.N. (1974). *Surf. Sci.* **41**, 113.

Jerrard, R.J., Ueba, H., and Davison, S.G. (1981). *Phys. Stat. Sol. (b)* **103**, 353.

Joannopoulos, J.D. and Cohen, M.L. (1975). *Phys. Rev. B* **10**, 5075.

Jones, H. (1962). *The Theory of Brillouin Zones and Electronic States in Crystals.* North-Holland, Amsterdam.

Jones, R.O. (1968). *Phys. Rev. Lett.* **20**, 992.

_______ (1972). *J. Phys. C* **5**, 1615.

_______ (1975). In *Surface Physics of Semiconductors and Phosphors.* (eds. C.G. Scott and C.E. Reed). Academic Press, London.

Jones, R.O., Jennings, P.J., and Jepsen, O. (1984). *Phys. Rev. B* **29**, 6474.

Kalkstein, D. and Soven, P. (1971). *Surf. Sci.* **26**, 85.

Knapp, J.A., Himpsel, F.J., and Eastman, D.E. (1979). *Phys. Rev. B* **19**, 4952.

Kolář, M. and Bartoš, I. (1973). *Czech. J. Phys. B* **23**, 179.

Kolář, M. and Davison, S.G. (1985). *J. Vac. Sci. Tech. A* **3**, 392.

Koster, G.F. and Slater, J.C. (1954a). *Phys. Rev.* **95**, 1167.

_______ (1954b). *Phys. Rev.* **96**, 1208.

Koutecký, J. (1957). *Phys. Rev.* **108**, 13.

_______ (1958). *Czech. J. Phys.* **8**, 148.

_______ (1960). *J. Phys. Chem. Solids* **14**, 233.

_______ (1961). *Czech. J. Phys. B* **11**, 565.

_______ (1965). *Adv. Chem. Phys.* **9**, 85.

_______ (1976). *Prog. Surf. Membrane Sci.* **11**, 1.

Koutecký, J. and Davison, S.G. (1968). *Int. J. Quantum Chem.* **2**, 73.

Koutecký, J. and Tomášek, M. (1960a). *J. Phys. Chem. Solids* **14**, 241.

_______ (1960b). *Phys. Rev.* **120**, 1212.

Kronig, R. de L. and Penney, W.G. (1931). *Proc. Roy. Soc.* **A130**, 499.

Landau, L.D. and Lifshitz, E.M. (1965). *Quantum Mechanics.* Pergamon, Oxford.

Levine, J.D. (1968). *Phys. Rev.* **171**, 701.

Levine, J.D. and Davison, S.G. (1968). *Phys. Rev.* **174**, 911.

Levine, J.D. and Freeman, S. (1970). *Phys. Rev. B* **2**, 3255.

Levine, J.D. and Mark, P. (1966). *Phys. Rev.* **144**, 751.

Levine, R.D. and Amos, A.T. (1967). *Phys. Stat. Sol.* **19**, 587.

Levy, R.A. (1968). *Solid State Physics.* Academic Press, New York.

Lifshitz, I.M. (1947). *Zh. Eksp. Teor. Fiz.* **17**, 1017, 1076.

Lippmann, B.A. (1957). *Ann. Phys.* **2**, 16.

Loucks, T.L. (1965). *Phys. Rev. Lett.* **14**, 693.

Löudon, R. (1959). *Am. J. Phys.* **27**, 649.

Löwdin, P.-O. (1951). *J. Chem. Phys.* **19**, 1396.

_______ (1962a). *J. Appl. Phys.* (Suppl.) **33**, 251.

_______ (1962b). *J. Math. Phys.* **3**, 969.

_______ (1963). *J. Mol. Spect.* **10**, 12.

_______ (1964). *J. Mol. Spect.* **14**, 119.

Lukes, T. (1969). In *Solid State Theory.* (ed. P.T. Lansberg). Wiley, London.

Lüth, H. (1975). *Appl. Phys.* **8**, 1.

MacColl, L.A. (1939). *Phys. Rev.* **56**, 699.

McLachlan, N.W. (1951). *Theory and Application of Mathieu Functions.* Oxford University Press, New York.

McRae, E. G. and Caldwell, C.W. (1976). *Surf. Sci.* **57**, 77.

McRae, E.G. and Kane, M.L. (1981). *Surf. Sci.* **108**, 435.

McRae, E.G., Landwehr, J.M., and Caldwell, C.W. (1977). *Phys. Rev. Lett.* **38**, 1422.

Many, A., Goldstein, Y. and Grover, N.B. (1965). *Semiconductor Surfaces.* North-Holland, Amsterdam.

Margenau, H. and Murphy, G.M. (1959). *The Mathematics of Physics and Chemistry.* Van Nostrand, Princeton.

Mark, P. (1970). In *Clean Surfaces.* (ed. G. Goldfinger). Dekker, New York.

Mathieu, É. (1868). *Jour. de Math. Pures et Appliquées* (Jour. de Liouville) **13**, 137.

Mattheiss, L.F. (1965). *Phys. Rev.* **139**, A1893.

Maue, A.W. (1935). *Z. Physik* **94**, 717.

Merzbacher, E. (1970). *Quantum Mechanics.* Wiley, New York.

Messiah, A. (1962). *Quantum Mechanics.* Vol. 1. North-Holland, Amsterdam.

Meyer, F. (1974). *Phys. Rev. B* **9**, 3622.

Meyer, F., de Kluizenaar, E.E., and Bootsma, G.A. (1971). *Surf. Sci.* **27**, 88.

Meyer, F. and Kroes, A. (1975). *Surf. Sci.* **47**, 124.

Meyerhof, W.E. (1947). *Phys. Rev.* **71**, 727.

Milanovic, V. and Tjapkin, D. (1982). *Phys. Stat. Sol. (b)* **110**, 687.

Miller, D.J., Heron, D.L., and Haneman, D. (1972). *J. Vac. Sci. Tech.* **9**, 906.

Modinos, A. (1974). *Surf, Sci.* **41**, 425.

_______ (1984). *Thermionic and Secondary Electron Emission Spectroscopy.* Plenum Press, New York.

Modinos, A. and Nicolaou, N. (1976). *Phys. Rev. B* **13**, 1536.

Mönch, W. (1973). In *Festkörperprobleme (Adv. Solid State Phys.)* **13**, 241.

Morgan, A.E. (1974). *Surf. Sci.* **43**, 150.

Morse, P.M. (1930). *Phys. Rev.* **35**, 1310.

Morse, P.M. and Feshbach, H. (1953). *Methods of Theoretical Physics.* McGraw-Hill, New York.

Mott, N.F. and Jones, H. (1958). *The Theory of the Properties of Metals and Alloys.* Dover, New York.

Müller, E.W. and Tsong, T.T. (1969). *Field Ion Microscopy, Principles and Applications.* Elsevier, Amsterdam.

_______ (1973). *Prog. Surf. Sci.* **4**, 1.

Müller, W. and Mönch, W. (1971). *Phys. Rev. Lett.* **27**, 250.

Nicolls, J.M., Mårtensson, P., and Hansson, G.V. (1985). *Phys. Rev. Lett.* **54**, 2363.

Nicolaou, N. and Modinos, A. (1975). *Phys. Rev. B* **11**, 3687.

_______ (1976). *Surf. Sci.* **60**, 527.

Nordheim, L. (1928). *Proc. Roy. Soc. A* **121**, 626.

Northrup, J.E. and Cohen, M.L. (1983). *Phys. Rev. Lett.* **49**, 1349.

Ohno, H., Mendez, E.E., Brum, J.A., Hong, J.M., Agullo-Rueda, F., Chang, L.L., and Esaki, L. (1990). *Phys. Rev. Lett.* **64**, 2555.

Olmstead, M.A. and Amer, N.M. (1984). *Phys. Rev. B* **29**, 7048.

Pandey, K.C. (1981). *Phys. Rev. Lett.* **47**, 1913.

_______ (1982). *Phys. Rev. Lett.* **49**, 223.

_______ (1982). *Physica B* **117/118**, 761.

_______ (1983). *J. Vac. Sci. Tech. A* **1**, 1099.

Pandey, K.C. and Phillips, J.C., (1974). *Phys. Rev. Lett.* **32**, 1433.

_______ (1975). *Phys. Rev. Lett.* **34**, 1451.

Peierls, R. (1930). *Ann. Phys., Lpz.* (5) **4**, 121.

Pendry, J.B. and Gurman, S.J. (1975). *Surf. Sci.* **49**, 87.

Penn, D. (1962). *Phys. Rev.* **128**, 2093.

Phariseau, P. (1960). *Physica* **26**, 737.

_______ (1961). *Physica* **27**, 590.

Phillips, J.C. (1969). *Covalent Bonding in Crystals, Molecules and Polymers.* Chicago University Press, Chicago.

Plummer, E.W. and Bell, A.E. (1972). *J. Vac. Sci. Tech.* **9**, 583.

Plummer, E.W. and Eberhardt, W. (1979). *Phys. Rev. B* **20**, 1444.

Plummer, E.W. and Gadzuk, J.W. (1970). *Phys. Rev. Lett.* **25**, 1493.

Priestley, H.A. (1985). *Introduction to Complex Analysis.* Oxford University Press, New York.

Pugh, D. (1964). *Phys. Rev. Lett.* **12**, 390.

Raimes, S. (1967). *The Wave Mechanics of Electrons in Metals.* North-Holland, Amsterdam.

_______ (1972). *Many-Electron Theory.* North-Holland, Amsterdam.

Read, M.N. (1986). *Phys. Rev. B* **32**, 2677.

Redondo, A., Goddard III, W.A., and McGill, T.C. (1982). *J. Vac. Sci. Tech.* **21**, 649.

Reihl, B., Frank, K.H., and Schlittler, R.R. (1987). *Phys. Rev. B* **30**, 7328.

Rijanov, S. (1934). *Z. Physik* **89**, 806.

Rowe, J.E. and Ibach, H. (1973). *Phys. Rev. Lett.* **31**, 102.

Rowe, J.E., Traum, M.M., and Smith, N.V. (1974). *Phys. Rev. Lett.* **33**, 1333.

Rowe, J.E., Christman, S.B., and Margaritondo, G. (1975). *Phys. Rev. Lett.* **35**, 1471.

Rundgren, J. and Malström, G. (1977). *J. Phys. C* **10**, 4671.

Rutherford, D.E. (1947). *Proc. Roy. Soc. (Edinburgh)*. **62A**, 229.

Salem, L. (1966). *The Molecular Orbital Theory of Conjugated Systems.* Benjamin, New York.

Saxon, D.S. and Hutner, R.A. (1949). *Philips Res. Rep.* **4**, 81.

Scheer, J.J. and Van Laar, J. (1963). *Phys. Lett.* **3**, 246.

Schlier, R.E. and Farnsworth, H.E. (1959). *J. Chem. Phys.* **30**, 917.

Schlüter, M., Chelikowsky, J.R., Louie, S.G., and Cohen, M.L. (1975). *Phys. Rev. Lett.* **34**, 1385.

Segall, B. (1962). *Phys. Rev.* **125**, 109.

Shikin, V.B. (1970). *Zh. Eksp. Teor. Fiz.* **58**, 1748 [Sov. Phys.-JETP **31**, 936].

_______ (1971). *Zh. Eksp. Teor. Fiz.* **60**, 713 [Sov. Phys.-JETP **33**, 387].

Shockley, W. (1939). *Phys. Rev.* **56**, 317.

Shockley, W. and Pearson, G.L. (1948). *Phys. Rev.* **74**, 232.

Slater, J.C. (1934). *Phys. Rev.* **45**, 794.

_______ (1952). *Phys. Rev.* **87**, 807.

Smith, N.V. (1985). *Phys. Rev. B* **32**, 3549.

Sokolov, A.A. (1934). *Z. Physik.* **89**, 806.

Spicer, W.E. (1966). In *Colloquium on the Optical Properties and Electronic Structure of Metals and Alloys, Paris, 1965.* (ed. F.A. Abelés). North-Holland, Amsterdam.

Srinivasan, U. and Davison, S.G. (1973). *Physica* **64**, 114.

Statz, H. (1950). *Z. Naturforsch.* **5a**, 534.

Steinmann, W. (1989). *Appl. Phys. A* **49**, 365.

Stephenson, P.C. and Bullett, D.W. (1984). *Surf. Sci.* **139**, 1.

Stęślicka, M. (1966). *Acta Phys. Polon.* **30**, 883.

_______ (1970a). *Physica* **45**, 506.

_______ (1970b). *Surf. Sci.* **19**, 318.

_______ (1974). *Prog. Surf. Sci.* **5**, 157.

_______ (1976). *Phys. Lett.* **57A**, 255.

_______ (1977). *Acta Univ. Wratisl.* **380**, 79.

Stęślicka, M. and Davison, S.G. (1970). *Phys. Rev. B* **1**, 1858.

Stęślicka, M., Kucharczyk, R., and Glasser, M.L. (1990). *Phys. Rev. B* **42**, 1458.

Stęślicka, M. and Perkal, Z. (1977). *Surf. Sci.* **62**, 406.

_______ (1980). *Solid State Comm.* **35**, 349.

Stęślicka, M. and Wojciechowski, K.F. (1966). *Physica* **32**, 1274.

Stęślicka, M., Zagórski, M., and Jurczyszyn, L. (1988). *Surf. Sci.* **200**, 209.

Straub, O. and Himpsel, F.J. (1984). *Phys. Rev. Lett.* **52**, 1922.

Streitweiser, Jr., A. (1966). *Molecular Orbital Theory for Organic Chemists.* Wiley, New York.

Strutt, M.J.O. (1928). *Ann. Physik* **85**, 129.

Subramanian, R. and Bhagwat, K.V. (1972). *J. Phys. C* **5**, 798.

Swanson, L.W. and Crouser, L.C. (1966). *Phys. Rev. Lett.* **16**, 389.

_______ (1967a). *Phys. Rev. Lett.* **19**, 1179.

_______ (1967b). *Phys. Rev.* **163**, 622.

Taloni, A. and Haneman, D. (1968). *Surf. Sci.* **10**, 215.

Tamm, I.E. (1932a). *Phys. Z. Sowjet.* **1**, 733.

_______ (1932b). *Z. Physik* **76**, 849.

Tamura, E. and Feder, R. (1990). *Z. Phys. B* **81**, 425.

Taylor, P.L. (1970). *A Quantum Approach to the Solid State.* Prentice-Hall, Englewood Cliffs.

Thakkar, A.J. and Stęślicka, M., (1978). *Surf. Sci.* **74**, 168.

Thörner, G. and Borstel, G. (1980). *Appl. Phys.* **41**, 99.

Tomášek, M. (1966a). *Phys. Stat. Sol.* **16**, 389.

_______ (1966b). *Surf. Sci.* **4**, 471.

Tomášek, M. and Koutecký, J. (1969). *Int. J. Quantum Chem.* **3**, 249.

Tomášek, M. and Pick, S. (1978). *Phys. Stat. Sol. (b)* **89**, 11.

Tomášek, M. and Šrubková, L. (1977). *Proc. 7th Intern. Vac. Congress and 3rd Intern. Conf. Solid Surfaces, Vienna, Austrian Vacuum Society*, p. 521.

Tsuji, M., Huzinaga, S., and Hasino, T. (1960). *Rev. Mod. Phys.* **32**, 425.

Ueba, H. and Davison, S.G. (1980). *J. Phys. C* **13**, 1175.

Uhrberg, R.I.G., Hansson, G.V., Nicholls, J.M., and Flodström, S.A. (1982). *Phys. Rev. Lett.* **48**, 1032.

Uhrberg, R.I.G., Hansson, G.V., Karlsson, U.O., Nicholls, J.M., Persson, P.E.S., Flodström, S.A., Engelhardt, R., and Koch, E.-E. (1984). *Phys. Rev. Lett.* **52**, 2265.

Van Laar, J. and Scheer, J.J. (1965). *Surf. Sci.* **3**, 189.

Van der Pol, B. and Strutt, M.J.O. (1928). *Phil. Mag.* **5**, 18.

Velický, B. and Bartoš, I. (1971). *J. Phys. C* **4**, L104.

Wagner, L.F. and Spicer, W.E. (1972). *Phys. Rev. Lett.* **28**, 1381.

Wannier, G.H. (1959). *Elements of Solid State Theory.* Cambridge University Press, London.

Weinert, M., Hulbert, S.L., and Johnson, P.D. (1985). *Phys. Rev. Lett.* **55**, 2055.

Whittaker, E.T. (1904). *Bull. Am. Math. Soc.* **10**, 125.

Whittaker, E.T. and Watson, G.N. (1958). *A Course of Mathematical Analysis*, 4th ed. Cambridge University Press, Cambridge.

Wigner, E. and Seitz, F. (1933). *Phys. Rev.* **43**, 804.

_______ (1934). *Phys. Rev.* **46**, 509.

Wilson, A.H. (1958). *The Theory of Metals.* Cambridge University Press, London.

Woodruff, D.P. and Smith, N.V. (1982). *Phys. Rev. Lett.* **48**, 283.

Yndurain, F. and Elices, M. (1972). *Surf. Sci.* **29**, 540.

Zangwill, A. (1988). *Physics at Surfaces.* Cambridge University Press, Cambridge.

Zak, J. (1984). *Phys. Lett.* **106A**, 399.

Ziman, J.M. (1965). *Principles of the Theory of Solids.* Cambridge University Press, Cambridge.

AUTHOR INDEX

SUBJECT INDEX

The manufacturer's authorised representative in the EU for product safety is Oxford University Press España S.A. of El Parque Empresarial San Fernando de Henares, Avenida de Castilla, 2 - 28830 Madrid (www.oup.es/en or product.safety@oup.com). OUP España S.A. also acts as importer into Spain of products made by the manufacturer.
Printed and bound by CPI Group (UK) Ltd, Croydon, CR0 4YY
14/09/2026
02216293-0008